S0-BID-791

The Human Genome Project in College Curriculum

The Human Genome Project in College Curriculum

Ethical Issues and Practical Strategies

Aine Donovan and Ronald M. Green
editors

Dartmouth College Press
Hanover, New Hampshire

PUBLISHED BY UNIVERSITY PRESS OF NEW ENGLAND
HANOVER AND LONDON

Dartmouth College Press
Published by University Press of New England,
One Court Street, Lebanon, NH 03766
www.upne.com

Printed in the United States of America
5 4 3 2 1

Library of Congress Cataloging-in-Publication Data
The human genome project in college curriculum : ethical issues and practical strategies / Aine Donovan and Ronald M. Green, editors.
p. cm.
Includes index.
ISBN-13: 978-1-58465-695-1 (cloth : alk. paper)
ISBN-10: 1-58465-695-6 (cloth : alk. paper)
1. Human Genome Project. 2. Human Genome Project—Study and teaching (Higher) I. Donovan, Aine. II. Green, Ronald Michael.
QH445.2.H92 2008
174'.957—dc22 2008001420

University Press of New England is a member of the Green Press Initiative. The paper used in this book meets their minimum requirement for recycled paper.

To the dedicated teacher-scholars who manage to take the complex issues that emerge in bioethics and make them understandable, relevant, and meaningful. And to the emerging teacher-scholars who will face an unlimited number of new bioethical challenges, especially Emily.

Contents

The Human Genome Project in College Curriculum

Introduction

Aine Donovan and Ronald M. Green
DARTMOUTH COLLEGE

The announcement in June 2003 that the first draft sequence of the human genome had been completed was a watershed moment in human history. For the first time, our species had at its disposal the instruction book for its own biological makeup. Ahead lay unprecedented new discoveries relating to the identification and diagnosis of disease, the development of new medical treatments, the understanding of human origins, the shaping of our genetic inheritance, and the control of human behavior. It is not an exaggeration to say that future historians may come to mark human history in terms of the pre- and post-genome eras.

The Genome Project was a thirteen-year-long effort to identify the sequence of three billion paired genetic letters, or "nucleotides," that form the instruction program for every feature of our biology. The final cost of the project was nearly $1 per for each pair of DNA letters. This completed sequence was published on the Internet along with the identification of an estimated 25,000 genes, those portions of the sequence coding for the proteins that form our cells and tissues. It is a sign of how rapidly the science and technology are moving that as a result of breakthroughs spurred by the Human Genome Project (HGP), the cost of DNA sequencing is now in free-fall. Genome researchers are already discussing a staggering new service: the "$1,000 genome." Many believe that by 2011, a small desktop unit in every physician's office might make it possible to rapidly discover the sequence variations that make each of us an individual. With this comes a vastly increased ability to find genetic factors that contribute to diseases. We will enter the era of personalized genetic medicine.

To some extent this is already happening. In a space of just two weeks in spring 2007, different groups of researchers, using information about the human genome developed by the HGP, announced the discovery of seven new genes associated with adult-onset diabetes and a common genetic variation that increases the risk of heart disease up to 60 percent in people of European descent. Although therapies based on these findings are not yet available, it is only a matter of time until interventions aimed at our genes become a major component of medical care.

Almost from its start, the Human Genome Project was recognized as raising unprecedented ethical questions for human society. In 1989, this prompted James Watson, the first director of the project and the co-discoverer of DNA, to propose to Congress that a percentage of federal support for genomic research be devoted to the study of these questions. This led to the establishment in 1990 of a National Institutes of Health (NIH) program to foster basic and applied research on the ethical, legal, and social implications (ELSI) of genetic and genomic research for individuals, families, and communities. The ELSI Research Program, as it came to be known, funds and manages studies, and supports workshops, research consortia, and policy conferences related to these topics.

During the first years of the ELSI program, attention focused on several key issues raised by genomic information and research. One was privacy. Who should have access to a person's DNA? The physician? Other family members? Government? Employers? And if any of these (or others) should have access, under what circumstances and with what constraints? Related to this is the question of discrimination. Is it ever permissible for decisions about someone's life to be made on the basis of his or her DNA? Employers and insurers currently take some medical conditions into account in their decision making, but should this extend to an individual's basic genetic makeup? If we allow this to happen, will we risk creating a class of permanent "genetic pariahs"? Lest the answers to these questions seem obvious, it is worth recalling that one's family's medical history has long been a factor in health and life insurance decision making. The challenge is to understand in what ways genetic information based on molecular DNA tests may be similar or different.

Another issue of great concern during the early years of genomic research and the ELSI program was patenting. When, if ever, should discoveries in the realm of pure or applied genome science come under the protections—and exclusive monopoly rights—accorded by patent law? The answer to this question embraces fields as diverse as constitutional and patent law, public policy, and ethics.

There is a series of complex questions that arise in connection with genomic research and the application of genetic knowledge to medical care, including the specific issues raise by prenatal genetic testing. The requirement of informed consent is fundamental to biomedical research and therapy. But since DNA is shared with close family members, and sometimes even with distant kin, what does informed consent mean in this context? Who is the subject that must consent to a research protocol or testing program? The individual seeking testing? Other family members whose privacy and security may be threatened by the disclosure of a relative's shared DNA status? Other members of one's ethnic group who may carry the same genetic traits that invite discrimination or stigmatization?

These and a host of related questions furnished an ample ELSI agenda during the 1990s. More recently, a further series of complex challenges has arisen for genomic research and therapy. Among the issues that have moved to the fore are the special ethical, legal, and social questions that come up as scientists and medical professionals attempt to translate the burgeoning body of genomic information to clinical health care. When is a technology, be it improved diagnostic methods or gene therapy, ready for clinical use? When does the premature implementation of a technology threaten important medical or social values? On a different front, what are the implications of using genetic information in non-health care settings? These include some familiar areas like employment and insurance, but with the progress of genomic science these concerns have more recently reached into such matters as education, adoption, criminal justice, and civil litigation.

Although behavioral genetics remains one of the least advanced areas of genomic science, some advances have been made as genetic contributions to mental disorders, addictions, and mood and temperament are slowly being uncovered. Since human behavior is at the heart of all our moral thinking, the ELSI issues here are prodigious, ranging from whether genetics should ever provide excusing conditions for criminal behavior, to the extent to which we should ever seek to use genomic therapies to alter someone's personality.

Finally, as genomic science inevitably becomes global, a series of questions arise about how different individuals, cultures, and religious traditions view the boundaries for the use of genomic information. The issue here is not ethical relativism, since there are presumably culture-transcending norms that belong to science and research ethics. Rather, it is about understanding how differing cultural beliefs can shape groups' approaches to and assimilation of genomic information. Unless this understanding is well developed, misinformation and misunderstandings can severely impede the progress of genomic science. One illustration is the Human Genome Diversity Project of the early 1990s, which, beginning as a well-intentioned effort to collect and categorize samples of DNA from all over the world, especially from dwindling human subcommunities, came to be viewed by some groups as a threat to their integrity and continuance.

The Human Genome Project and the way we answer these questions promise to impact everyone, but the generations of young people being educated at the start of the twenty-first century will determine how the fruits of the HGP are used and whether they contribute to the ongoing improvement of human life or lead to new forms of coercion and injustice. As future medical researchers and clinicians, today's college students will shape the directions of genomic technologies. As financiers and investors, they will determine what new industries genomics creates. As legislators, politicians, and voters, they will craft the laws that govern the utilization of genomic biotechnology. As parents, they will be affected by our new powers of predictive genetic analysis. Will they use these powers to enhance the lives of their children, or will they create the repressive "brave new world" sketched by Aldous Huxley?

College and university teachers have a special role to play in communicating and interpreting the meaning of genomic discoveries. In the mid-1990s, Dartmouth College's Ethics Institute, working on the premise that today's college-age students will live their lives in a world shaped by genomic discoveries, applied to the National Institutes of Health for a grant to educate college and university teachers about the ethical, legal, and scientific implications (ELSI) of the HGP. We proposed a series of innovative summer faculty workshops for this purpose. We would gather a selected group of top college and university teachers at Dartmouth to examine some of the leading ELSI issues raised by the genome project. The workshop would provide some background on ELSI issues. The goal was to help spur development of courses all across the country that would introduce undergraduate students to the results of the HGP and the challenges posed by it.

Our core faculty consisted of a bioethicist and religion scholar (Ronald M. Green), an ethicist (Aine Donovan), a molecular biologist (David Bzik), and a lawyer (Albert Scherr). Supplementing these resources were other national figures such as Dean Hamer (behavioral genetics), Adrienne Asch (genetics and disabilities), Jennifer Axilbund and Wendy McKinnon (genetic counseling), Alice Wexler (genetics and women's studies), John Fletcher and Erik Parens (bioethics), Charmaine Royal and Rick Kittles (genetics and race), Jay Kohler (biostatistics), and Steven Holtzman (biotechnology and gene patenting).

Our grant proposal began an adventure that carried over a total of three grant cycles, grew enormously in scope, and eventually reached over two hundred and forty college and university teachers. Most of these participants were from the United States, but in an effort to expand our understanding of how societies will cope with the outpouring of new genomic information, we made sure to include international scholars in each offering of the program. By the final three-year funding cycle, demand for the workshop had increased so much that we were inspired to offer three week-long programs of twenty participants each. In this final cycle, we partnered with Howard University in Washington, D.C., both to help us meet the growing demand and to bring to bear the expertise of Howard's faculty and scientists on the increasingly important area of race and genetics.

The culminating event in this decade-long project was a conference in the summer of 2005. This was a scholarly event to which past participants were invited to submit papers reflecting the research and teaching innovations that resulted from their work in the summer institute. More than one hundred and twenty participants responded to this invitation, an extraordinary turnout that testified to the impact of this experience on them. Equally extraordinary were the quality of the papers and the intensity of conversation. Keeping in mind that many of these people had never met one another before—their involvement was spread over nine separate summers and many distinct workshops—they nevertheless constituted a community of scholars and scientists joined by their shared commitment to understanding and teaching the most cutting-edge issues raised by the genome project.

From the beginning we maintained three fundamental commitments. The first was an approach that might be called "disciplined interdisciplinarity." Comprehending ELSI issues requires the ability to reach across disciplinary lines. A geneticist can minutely describe the chromosomal abnormalities that contribute to Down's syndrome. But whether the currently vast expansion of prenatal screening for this disorder is a good idea—indeed, whether Down's Syndrome should even be seen as a disorder—can only be addressed by people bringing perspectives from such diverse fields as psychology, bioethics, and disability studies. ELSI issues thus require the ability to reach across disciplinary lines. At crucial points they also require disciplinary expertise. A philosopher cannot be expected to explain the statistical pitfalls in forensic genetics any more than a statistician or biologist can explain the justice considerations underlying constitutional principles. This commitment to disciplined interdisciplinarity also underlay our preference for two-person "teams" among the teachers invited to participate in the summer institute. We believe that, at a minimum, adequate ELSI teaching requires bifocal disciplinary vision. We also strived to provide faculty institute participants with the resources needed to achieve "disciplined interdisciplinarity" at their home institutions. This included a rich array of audiovisual materials resulting from the course and institute (special lectures or panels), as well as Internet materials, many of which are regularly updated on our web site (http://www.dartmouth.edu/~ethics/smi.html) on a regular basis.

Our second major commitment was to a case-based approach to the undergraduate teaching of ELSI issues. This utilizes selected genetic disease conditions, genetic initiatives, and episodes in the history of genetics as organizing tools for exploring ethical, legal, and social issues of continuing importance. Case modules focused on carrier screening for cystic fibrosis, testing for BRCA mutations, and the eugenic movement as an episode in the history of genetic science. In each instance, the "case" was organized around a challenging question or problem. Should we extend screening for cystic fibrosis carrier status to all young people, as opposed to the limited testing now conducted only on likely carrier couples? Should information about the genetic basis of behaviors be admitted to the courtroom as an exonerating or mitigating factor in sentencing decisions? This case-based approach to teaching becomes even more important as genomic science advances. It is impossible for anyone to communicate or comprehend all of the information being produced by genomic research. Today's knowledge is rapidly rendered irrelevant by tomorrow's discoveries and challenges. But what students can and should learn to do how to learn. By conveying the array of skills needed to understand and address a complex problem, a case-based approach models the skills that students will need to bring to new and unanticipated developments.

Our third commitment was to incorporate the perspectives of women and minorities in the curriculum and among our summer institute faculty and participants. Human genetics has powerful and differential implications for women, members of minority groups, and individuals with disabilities. As the HGP moves forward to the study of the role of genes in human variation, it would be irresponsible not to ensure an active presence of individuals from diverse backgrounds in the study and discussion of ELSI issues. In addition to beginning each program by dwelling on the history of the misuse of genetic information under the guise of "eugenics," we emphasized the impact of genetic testing and screening programs on disabled people and other marginalized social groups. Our collaboration with Howard University during the last three years of the program put us in direct contact with cutting edge issues in race and genetics. As we were meeting, Howard researchers were testifying before the FDA on the appropriateness of targeting medications to black Americans on the basis of widely shared (but by no means universal) genetic susceptibilities.

The chapters in this volume represent the thinking, experience, and creativity of the college teachers who participated in over a decade of summer institutes. The chapters are not easily categorized because they reflect the commitment to interdisciplinarity that was a hallmark of this effort. Each chapter speaks to some particular facet of the ELSI agenda, but is also informed by a wider array of concerns. Together, they provide an extraordinary starting point for anyone interested in bringing genomics into a liberal arts education.

Some of the very basic value questions raised by genomic advances are explored in these chapters. Students enter the classroom asking how genomics will impinge on some of their most cherished institutions and values. For many people around the world and in this country, genetic knowledge threatens to disorder our relationship to the divine. "We should not play God," is the warning voiced in numerous popular and scholarly treatments. Without even realizing it, undergraduates in particular resonate to these concerns. But what exactly is the sphere of human authority, and does it extend to the control of genetic information and power? The chapters explore the many particular facets of the larger questions of "what does it mean to be human?" and "who decides, and when, what it means to be perfect?" For example, when does genetic screening become a practice akin to eugenics? Should the medical community ever require a parent to "correct" a genetic flaw? The chapter by Sonja Eubanks begins by focusing on the anguished decision making that faces parents when asked to consider fetal surgery to correct a genetic defect in their expected child. In this experimental setting, parents, counselors, and physicians seeking a healthy birth must work together to assess the risks to the survival of the child and the mother. Still more distant on the horizon is the use of gene therapy to correct genetic defects at the earliest possible stage. Maureen Sander-Staudt examines the work of bioethicist Sara Goering for its implications in helping us decide which conditions are appropriately regarded as medical issues worthy of correction and which lie in a less urgent, and maybe even prohibited, area of gene enhancement. The practice of pre-implantation genetic diagnosis (PGD) is explored as a potentially manipulative means of using one sibling for another's medical purposes. Drawing on Jodi Picoult's novel *My Sister's Keeper*, Terrance McConnell critically examines the value choices that these new technologies pose. Students who have read the novel are plunged into a world of moral uncertainty when they enter the dilemma of trying to save a child's life by creating a "savior sibling." The use of the novel is an especially relevant means of highlighting the experiential component of the HGP.

On a more practical level, genomics has powerful implications for people's access to health and life insurance, potentially giving applicants or insurers privileged access to private information that can greatly affect a person's health or employment situation. Ben Eggleston argues that prohibiting employers' or insurers' access to such information is misguided, arguing that the emphasis instead should be placed on reforms in our nation's health care system. Continuing this emphasis on privacy, law school educator Julia Gladstone maintains that understanding the relationship of privacy and autonomy and identity and DNA is fundamental to developing a fair system of privacy protections.

Biologist Alexander Werth evidences the fruitfulness of our interdisciplinary conversations by bringing out of his work on the genome project its surprising implications for the teaching of modern biology. According to Werth, one of the greatest values of genomic information is the way it enhances our understanding of genetic variation within our species and in relation to other species. The teaching of evolutionary biology, Werth suggests, can be transformed by attention to modern genomics.

While every chapter in this book has an emphasis on teaching the human genome project, the chapters gathered in the second half of the volume explicitly address pedagogical matters. Few college courses inspire as much heated debate as courses dealing with the HGP. The subject matter is fraught with moral complexity that is not easily categorized or accepted by society. College students are a microcosm of the society from which they come: They bring their religion, politics, prejudices, ideals, and ignorance into the classroom. The educational objective of teaching a class that has the HGP as a focal point is to understand those complexities, but not necessarily to offer solutions to the ambiguities. These chapters tease out just some of the tensions and offer some tested methods for teaching.

Tom Segady explores the often hidden "structures of power" that will shape the way we understand and respond to genomic discoveries. The increasingly self-described religious orientation of college students poses a unique challenge for many college professors. Students who view science through a religious lens can feel threatened by a subject matter that may appear to contradict fundamental beliefs. Working in the context of a religiously affiliated college, Donna Yarri and Spencer Stober show how a multidisciplinary approach (religion–biology) can help students incorporate genomic knowledge into their spiritual concerns. Ronald Green and Aine Donovan also address the religious elements of the HGP by exposing the deep religious motifs found in the film *Gattaca*. This film has been widely—and successfully—used in bioethics courses as a jumping off point for the discussion of genetic discrimination. But *Gattaca* is rarely examined for its pervasive religious symbolism and vision.

Bringing the legal issues of the HGP into an undergraduate class poses problems of depth, scope, and size. Yet the legal implications are a primary concern for anyone who strives to offer a complete overview of the issues. With a focus on the implications of behavioral genetics for issues of punishment, Erica Beecher-Monas asks explicitly how genomics can shape the teaching of criminal and evidence law. Another approach for integrating the varied strands of the HGP into one class is found in the chapter by Bethany Hicok and Joshua Corrette-Bennett. They argue for a "cluster approach" in which several separate courses are linked for teaching in an interdisciplinary mode. The authors present the advantages of an interlinked integrated program that not only provides students with expanded scholarly vision but also brings faculty together in a collaborative effort. This focus on faculty is echoed in subsequent chapters by arguing for a reasoned approach toward the subject matter and, most importantly, for eliminating "fanatics" from the bully pulpit—whether for or against technology. For example, at the heart of genomic debates, there is the question of nature versus nurture. Using this issue as an introduction to issues in the history of science, Myles Jackson insists that scientists should not allow the public debate on ELSI to be carried out by anti-science fanatics. He adds that it would also be unwise for humanities students to surrender ELSI discussions to scientists alone. Finally, Anne Galbraith concludes the volume by addressing a persistent problem for college science teachers: how to cover the necessary science information while also alerting students to the humanities and social science aspects of science research. She believes that attention to the social and ethical ramifications of the genome project can actually add to building of an adequate science foundation.

We are at the beginning of what may come to be regarded as the genomic century. The scientists and scholars of the Dartmouth summer institutes on the ethical, legal, and social implications of the human genome project initiated a conversation that reverberates in classrooms, courtrooms, laboratories, and public policy forums around the country. We know that the sophistication of ELSI discussion will grow enormously in the years ahead. Recent surveys indicate that college students have the most sophisticated scientific/technological knowledge of any generation, but what they lack is the wisdom to put that knowledge to good use. That is the job of college professors, from every discipline and from every region of the country. This volume offers some insight that will advance the proper use of the knowledge about the Human Genome Project.

Ethical Dilemmas in Clinical versus Research-Based Medicine

Fetal Surgery for Neural Tube Defects

Sonja Eubanks

UNIVERSITY OF NORTH CAROLINA–GREENSBORO

Options for medical therapies have progressed dramatically in recent years. However, not all procedures are available on a clinical basis and there are differences between clinical and experimental therapeutic options. Extensive trials are generally necessary to establish the efficacy of a clinically accepted procedure. Clinical medicine is focused on the care of an individual. Clinical research is designed to answer a particular scientific question for use as general knowledge with future patients (Miller and Brody 2003). Both involve various treatment regimens or procedures. It is often difficult for patients to understand the differences between clinically available services and those offered as part of a research study. A patient may assume that a procedure has been proven to be safe and effective if it is offered in a medical center, when in fact the procedure may not be in that patient's best interests. This confusion has been termed the "therapeutic misconception" (Jansen 2006; Miller and Brody 2003). Even if patients understand that a procedure has not been proven safe and effective, they still may be willing to participate in the research because it is the only means by which the particular procedure or treatment is available. The patient may face the dilemma to either participate in research in order to be able to receive new and possibly better services or to avoid risks and accept currently available clinical services. One example of a therapeutic dilemma that a patient may face is the option for participation in experimental fetal surgery for neural tube defects. This essay presents information about this research study and the ethical questions that arise for patients, health care providers involved in the care of pregnant women carrying a fetus with a neural tube defect, and researchers.

Neural tube defects (NTDs) are common birth defects, occurring in 1 or 2 of every 1,000 live-born babies in the United States. NTDs occur when the neural tube, or spine, fails to close properly at approximately twenty-eight days of gestation. The term *spina bifida* is used specifically when referring to openings of the spine rather than the skull. *Myelomeningocele* is another term for a specific type of opening in the spine, where the spinal cord protrudes through the opening. Larger and higher openings of the spine are associated with more severe symptoms than smaller, lower openings. People with the more severe forms are not able to walk or control their bowel or bladder functions and may require a shunt placement and suffer with clubbed feet. There can also be seizures and mental retardation if the hydrocephalus is not successfully treated. People with the less severe forms have some but not all of these symptoms. People with spina bifida have symptoms that are lifelong and affect their abilities to perform functions that many people take for granted. The cause for NTDs is not clear, but appears to be a combination of environmental and genetic factors in most cases. When a couple has had one child with an isolated NTD, they have an increased chance of about 3–5 percent to have another.

When a child is born with spina bifida, the typical treatment is surgery to close or cover the opening. This treatment does not prevent the symptoms from occurring. Ongoing care for individuals with spina bifida is based on symptoms, such as placing a ventriculo-peritoneal shunt when hydrocephalus is present. Many medical specialists, such as physical and occupational therapists, urologists, orthopedists, neurologists, and others, are involved in the care of a person with spina bifida. While symptoms of spina bifida are chronic and have a significant impact on a person's day-to-day functioning, spina bifida is typically not a life-threatening condition. Due to the lack of improvements in the care for those with spina bifida, physicians began to wonder whether one way to improve outcomes would be to try to cover the spine earlier, during pregnancy. It has been theorized that some of the damage that occurs to the spine in individuals with spina bifida occurs during pregnancy when the spinal cord is exposed to the amniotic fluid or by the hydrocephalus that is untreated until after birth. Would covering the spine earlier minimize the symptoms, give more people the ability to walk or control their bowel or bladder functions? Could it prevent the need for the treatment of hydrocephalus after birth? These questions and some attempts in animal models led physicians to begin performing in utero surgery to repair spina bifida on a research basis in 1997. The surgery was the first fetal surgery attempted for a non-life-threatening condition. Fetal surgery had been performed in the 1980s for life-threatening conditions such as diaphragmatic hernia with limited success (Couzin 2006). In the case of spina bifida, the fetal surgery was not about saving the life of the fetus, but about attempting to improve the quality of life after birth.

In the last few decades, the technology to detect NTDs during pregnancy has improved so that many cases are detected prenatally. One method for detection is a blood test performed on the mother at fifteen to twenty weeks of gestation, called the *maternal serum multiple marker screening*. This screening can detect about 80 percent of fetuses with an NTD. In addition, a detailed ultrasound performed by a trained technician or physician can detect approximately 90 percent of fetuses with NTDs at eighteen to twenty weeks of gestation. Amniocentesis, where a small portion of fluid from the amniotic sac is obtained by a needle, is a definitive test for NTDs.

From 1997 to 2003, women who were found to be carrying a fetus with an NTD could be referred to one of a few centers in the country that offered closure of the fetal spine in utero. The procedure was free of cost to patients because it was performed on a research basis. After discussing the procedure, women could decide whether or not they wanted to have the fetal surgery performed. The procedure was performed on more than two hundred fetuses and the development of the children has been followed over time. Data from these cases indicated that there was no noted improvement in the ability to walk or control bowel or bladder functions. However, there did appear to be an improvement in the need for shunt placement at birth. This was related to an improvement in the degree of hind-brain herniation into the spinal column. Some of those with the in utero repair still required shunt placement later. It is not clear whether this improvement may ultimately allow for an improvement in cognitive abilities (Bruner et al. 1999, 2004; Johnson et al. 2006).

The mothers and fetuses also experienced complications of the procedures. There was a significant increase in the rate of premature delivery. In one of the sites that performed 178 fetal surgeries, 12 percent of babies were delivered before thirty weeks of gestation (Bruner and Tulipan 2005). Prematurity brings its own set of risks to infants including an increased chance for cerebral palsy, blindness and death. In fact, 5 of the 178 fetuses at this site died due to complications of the surgery and prematurity. Fetal surgery also brings the risk of further trauma to the already damaged spine and an increased chance for leakage of amniotic fluid, which could impair the development of the fetal lungs. These complications in addition to the spina bifida create an even more complex medical situation. The complications for the mothers includes uterine rupture requiring the need for a complete hysterectomy in about 2 percent of cases, wound infections, bleeding requiring a transfusion, side effects of medications and anesthesia, and the requirement that future births occur by cesarean section (Golombeck et al. 2006). In 2001, the American College of Obstetricians and Gynecologists published a committee opinion entitled "Fetal Surgery for Open Neural Tube Defects" that stated that a multicenter, randomized, controlled trial should be performed before the fetal surgery would become a clinically available procedure. This would allow a more detailed, thorough, and long-term evaluation of the procedure by allowing a direct comparison of fetuses who do have the prenatal surgery with similar fetuses who do not.

In February 2003, a clinical trial funded by the National Institute of Child Health and Human Development began recruiting subjects. The trial is called the Management of Myelomenigocele Study or MOMS trial. The goal of this study is to recruit two hundred women carrying a fetus with spina bifida to participate in a randomized controlled trial. This means that all two hundred women will go through an informed consent process to learn about the surgery at one of three clinical centers. Of those who choose to participate in the trial, one hundred will actually have the surgery and one hundred will be assigned to a control group that does not get the surgery (Bruner and Tulipan 2005). The surgery is no longer available in the United States in any other center besides those participating in the MOMS trial. Therefore, while in the past all women who wanted the surgery could have access to it, currently, some women who are convinced that this is a vital procedure for their fetus will ultimately be assigned to the control group and not have access to the surgery. While the need for a controlled trial for fetal surgery for spina bifida is clear, it has resulted in frustration for some patients who feel their right to choose a treatment option for their fetus has been limited.

In addition, in order to get meaningful data, there are guidelines about who is eligible to participate in the MOMS trial. Participants must be eighteen or older, be a U.S. resident, have a fetus with spina bifida detected before twenty-five weeks gestation, have an amniocentesis procedure that shows the fetus has normal chromosomes, have a fetus with no other structural anomalies, be pregnant with only one fetus, and have no signs of preterm labor or maternal illnesses like diabetes. In addition, women who participate and are assigned to the control group will return home for the remainder of the pregnancy, but must be able to travel back to one of the study sites to deliver by cesarean section. Those assigned to the fetal surgery group will stay at the study site for the remainder of the pregnancy and deliver by cesarean section there (Management of Myelomeningocele Study 2003). Therefore, in order to be eligible to participate in the MOMS, a woman must be able to commit to staying at the research center for the remaining three to four months of her pregnancy. In exchange for study participation, patients will have follow-up evaluations and medical care for their child with spina bifida at no cost through the follow-up period of the study. The study is designed to allow complications for mother or baby from the procedure to be linked directly to the procedure rather than to any other factor. However, these necessary restrictions on eligibility mean that additional women who have a fetus with spina bifida will not even have a chance to be assigned to the surgery group.

The MOMS protocol does include a one-and-a-half-day education and informed consent process where women and couples meet with a variety of specialists including medical ethicists, social workers, clergy, nurses, perinatologists, neonatologists, pediatric neurosurgeons, anesthesiologists, a study coordinator, and a genetic counselor. This process is meant to help families make decisions they are comfortable with. However, the question arises of whether or not women in a time of crisis, such as finding out that the fetus they are carrying has a debilitating condition, are able to absorb this information and make a clearly formed decision. In my own work providing genetic counseling to some of these families, they indicate to me that if there is any possibility of improving the condition for the child, the parents feel they have to try it. Parents say things like, "What would we do if this study shows a great improvement for children and we didn't have it for our child? How could we live with that?" Parents are also in a situation of having to make a fairly quick decision in many cases, as the study is designed so that surgery is performed before twenty-five weeks of gestation. The typical time frame of finding out about this condition is eighteen to twenty-two weeks of gestation. Also, patients may hear about fetal surgery in the news and have an expectation that it is more successful than it truly is. The media often report on the patients for whom a procedure made a drastic change, rather than on those with the more typical result of a procedure.

There are other factors that affect a family's ability to make a clearly informed decision about participating in the MOMS. One of these factors is the possibility of impartiality of the study team in relaying information to potential study participants. Presumably, the researchers who perform the procedures are advocates of the study and want to recruit patients. In addition, how much does therapeutic misconception affect the informed consent process in this case? Do families really understand that there are significant risks associated with fetal surgery, above and beyond other types of surgery? Will families in such a vulnerable state also deal with therapeutic misestimation, where they overestimate possible benefits and underestimate possible risks because of their desire to help their child and deal with the guilt of bringing a child in to the world with a birth defect? It is already clear from discussions with various individuals involved in the counseling of women in the pre-MOMS era that women often come in determined to have the procedure, regardless of the risks described (Couzin 2006). Will these families understand that the surgery may not improve the outcome of the condition but may actually make the condition worse and adversely affect the mother's reproductive health?

If a woman completes the informed consent process and chooses to participate in the MOMS and is chosen to receive the fetal surgery, then the preparations for surgery begin. It may be helpful to provide a general description of the surgery here, as fetal surgery is a unique procedure. Before surgery, the mother is given steroids to help the fetal lungs develop in case of a preterm delivery. During the procedure, the mother is given anesthesia through an epidural, as well as general anesthesia that crosses the placenta and also reaches the fetus. The mother is given antibiotics intravenously and then an incision is made on her abdomen above the uterus. The uterus is then removed from the body through this incision and rests on the mother's abdomen. An incision is then made in the uterus so that the fetus can be directly manipulated. The amniotic fluid around the fetus is removed and stored and the fetus is placed in the uterus with the spine up. A repair of the fetal spine occurs at this point, in the same way that it would if performed after birth. Some extraordinary photographs of these procedures have been taken and published in the popular press, showing the fetal hand touching the hand of the surgeon. When the fetal repair is finished, the amniotic fluid is placed back in the uterus, the uterus is closed with a substance akin to glue, the uterus is placed back in the mother, and the mother's abdominal incision is closed. The mother then receives medications to prevent early labor and returns weekly to the clinic for monitoring (Bruner and Tulipan 2005).

Recruitment of patients to the MOMS has been slower than expected. It appears that not all obstetricians feel that this procedure is something they should be informing their patients about: that the risks do not outweigh the benefits at this point (Lyerly et al. 2001). In a clear call to obstetricians in the United States, those involved with the MOMS trial submitted personal narratives written by parents who previously had the fetal surgery and reported on their experiences in the journal *Clinical Obstetrics and Gynecology* (Gonzalez-Abreu 2005; Kennedy and Kennedy 2005; Williamson and Williamson 2005). There was a series of these narratives by parents who all had positive feelings about the decision-making process for fetal surgery. The researchers did not include any narratives by people who went through the informed consent process and decided not to participate in the fetal surgery. The study team seemed to be sending the message to obstetricians that they have an obligation to their patients to inform them of this study (Chescheir and D'Alton 2005). A majority of obstetricians believe that having the fetal surgery is not in the medical best interests of the mother and that there is no clear evidence of substantial benefit to the child (Lyerly et al. 2001). It will be interesting to follow the long-term outcomes of those with fetal surgery to determine if there is a real medical benefit. If so, will obstetricians be more willing to refer patients for fetal surgery in light of the significant chance for maternal and fetal complications? Will maternal and fetal complication rates decline as the researchers gain more experience with the procedure? In this time, when the answers to those questions are not known, should obstetricians refer women to the MOMS? Will obstetricians continue to feel their primary concern is to limit risks to the mother, or will the researchers convince them that they have an obligation to promote improved outcomes for the fetus (Chervanak and McCullough 2002)?

Families who have just discovered that their fetus has a significant medical condition are left to sort out a very complex decision about fetal surgery. The information presented is detailed and complex even for those with some medical background. There are a number of different ways that families may think about this decision. Families may feel that they have to choose the surgery, because that is what a responsible, loving parent would do. The parents may be dealing with their own guilt, feeling that they somehow caused the condition, and this is one way to make things better. They may feel that good parents would do everything they can to help their child, even if it means taking some risks. The parents may feel pressure from friends and family to participate in the MOMS because it would be cruel not to help the child if they could. The parents may feel either real or perceived pressure from the primary obstetrician not to have the surgery and from the researchers to have the surgery. They may feel that they can't say no, even in the face of risks to the mother and her future reproductive health and risks for other complications to the child (Parens 2006). Parents want to make a decision that is in the best interests of the child, but this is a case in which best interest is difficult to discern.

There is a complex interplay of the family, the primary obstetrician, and the research team that occurs in making a decision about fetal surgery for spina bifida. Is it reasonable or fair to ask families to make these kinds of decisions? If not, how will new procedures that have the possibility to improve outcomes be studied? The risks inherent in fetal surgery may seem more feasible when the fetus has no other chance to survive, but what about in cases where the fetus will most likely survive but with chronic symptoms? Are the risks of fetal surgery worth the possible benefits in cases of nonlethal conditions? Are risks to mothers like uterine rupture and future infertility and risks to children like death from prematurity too high to take in order to study a new procedure? These issues are also relevant to fetal surgery for other conditions and are not likely to abate as the new frontier in fetal surgery appears to involve attempts to correct various forms of congenital heart defects (Couzin 2006). Navigating this area of medicine is a challenge for families, health care providers, and researchers alike.

References

American College of Obstetricians and Gynecologists Committee Opinion. 2001. Fetal Surgery for Open Neural Tube Defects, Number 252: March.

Bruner, Joseph, and Noel Tulipan. 2005. Intrauterine Repair of Spina Bifida. *Clinical Obstetrics and Gynecology* 48 (4): 942–955.

Bruner, Joseph, et al. 1999. Fetal Surgery for Myelomeningocele and the Incidence of Shunt-Dependent Hydrocephalus. *JAMA* 282 (19): 1819–1825.

———. 2004. Intrauterine Repair of Spina Bifida: Preoperative Predictors of Shunt-Dependent Hydrocephalus. *American Journal of Obstetrics and Gynecology* 190: 1305–1312.

Chervenak, Frank, and Laurence McCullough. 2002. A Comprehensive Ethical Framework for Fetal Research and Its Application to Fetal Surgery for Spina Bifida. *American Journal of Obstetrics and Gynecology* 187: 10–14.

Chescheir, Nancy, and Mary D'Alton. 2005. Evidence-Based Medicine and Fetal Treatment: How to Get Involved. *Obstetrics and Gynecology* 106 (3): 610–613.

Couzin, Jennifer. 2006. Desperate Measures. *Science* 313 (August 18): 904–907.

Golombeck, Kirsten, et al. 2006. Maternal Morbidity After Maternal–Fetal Surgery. *American Journal of Obstetrics and Gynecology* 194: 834–896.

Gonzalez-Abreu, Emily. 2005. Parental Voices: For Angeline: A Mother's Reflection and Emotional Struggle with the Loss of a Typical Child and How She Stumbles Into an Unexpected World of Special Needs and the People Who Live There. *Clinical Obstetrics and Gynecology* 48 (3): 518–526.

Jansen, Lynn. 2006. The Problem with Optimism in Clinical Trials. *IRB: Ethics & Human Research* 28 (4): 13–19.

Johnson, Mark Paul, et al. 2006. Maternal–Fetal Surgery for Myelomeningocele: Neurodevelopmental Outcomes at 2 Years of Age. *American Journal of Obstetrics and Gynecology* 194: 1145–1152.

Kennedy, Dean, and Lesley Kennedy. 2005. Parental Voices: Our Journey with Grace. *Clinical Obstetrics and Gynecology* 48 (3): 534–539.

Lyerly, Anne, et al. 2001. Attitudes of Maternal–Fetal Specialists Concerning Maternal–Fetal Surgery. *American Journal of Obstetrics and Gynecology* 185: 1052–1058.

Management of Myelomeningocele Study (MOMS). 2003. Overview of MOMS. http://www.spinabifidamoms.com/english/overview.html.

Miller, Franklin, and Howard Brody. 2003. Therapeutic Misconception in the Ethics of Clinical Trials. *Hastings Center Report* 33 (3): 19–28.

Parens, Erik (ed.). 2006. *Surgically Shaping Children: Technology, Ethics andthe Pursuit of Normality*. Baltimore: Johns Hopkins University Press.

Williamson, Jason, and Susan Williamson. 2005. Parental Voices: The Positive Impact of Medical Professionals. *Clinical Obstetrics and Gynecology* 48 (3): 512–17.

Gene Therapy, *Gattaca*, and Sara Goering

A Procedure for Ethical Deliberation on Gene Selection

Maureen Sander-Staudt
ARIZONA STATE UNIVERSITY

Gattaca is a science fiction film that is every day becoming less than fictional. In the style of Huxley's *Brave New World*, *Gattaca* is set in a futuristic society where most children are conceived in test tubes and selected for their genetic superiority. Those with the misfortune of a "faith birth," like the protagonist Vincent, are marked as "in-valids" and "de-gene-rates," members of a vilified subcaste. Born with a weak heart, Vincent is an immediate disappointment to his father. Denied entry to space academy due to his low pedigree, he constructs an elaborate identity theft and rises from being a custodian at Gattaca Aerospace Corporation to Navigator First Class.

We as a society seem far from the day when the majority children are "designed" with precision and birthed through calculated mechanical means, but we are ever moving closer to this kind of world. Rising numbers of children are born through in vitro fertilization (IVF) technologies, and are being selected for and against on the basis of genetic features. People whose home countries have some ban against the selection of fetal gender (France, the United Kingdom, Germany, Switzerland, Austria, Japan, India, and the Netherlands) are traveling to the United States, where the procedure is currently legal and untracked (Johnson, 2006). It is standard practice for pregnant women to be offered tests to detect genetic information about fetuses, and most expecting parents hope for a "healthy, normal child." The birth of a child with a disease or disability can be financially devastating, as well as emotionally taxing. Many parents who would not abort a child with a genetic malady, or feel less love for a child born with a defect, might feel comfortable about selecting from a set of genetically prescreened blastocytes, or selecting against a particularly harmful genetic condition preimplantation, in utero, or postpartum. The current trend of paying dearly for eggs and sperm from individuals with certain desirable traits suggests a potential market for "enhancement" via gene splicing. It is not a stretch to speculate that market forces and technological advances in gene mapping and artificial reproduction will combine to generate questionable new eugenic practices similar to those portrayed in *Gattaca.*

But while some painful, disabling, or life-threatening genetic conditions seem to be legitimate candidates for gene therapy, others, such as those relating to beauty, intelligence, and personal character, are more problematic. The film *Gattaca* reflects the uneasiness we feel about genetic selection in the context of past eugenics movements. This film effectively shows the need to decide which uses of this technology are morally permissible and which are not. Here I describe how *Gattaca* can be used in conjunction with the essay written by Sara Goering entitled "Gene Therapies and the Pursuit of a Better Human" to help students of bioethics identify conditions that are morally legitimate candidates for gene therapy. I explain how Goering's essay offers a useful procedure for distinguishing between true maladies and social prejudices that can be used in a variety of pedagogical forms. I show the limits of this usefulness by explaining how genetic dispositions pertaining to moral character are particularly thorny issues not adequately addressed by either *Gattaca* or Goering. I also describe how more advanced students of bioethics can be prompted to dialectically build from Goering's argument by adopting positions rooted in the Ethics of Care and in Libertarianism.

Historical knowledge of the darker aspects of eugenics programs lends a grim realism to *Gattaca*, but a balanced use of this film for the ethical analysis of gene therapy requires also that we recognize the more nuanced aspects of eugenics. It can be argued that vaccinating children against disease, timing one's family to be earlier in life rather than later, taking prenatal vitamins, avoiding situations known to cause birth defects, and even being choosey about with whom one procreates are instances of eugenics. One of the many virtues of Goering's essay that makes it complementary to *Gattaca* is that it briefly highlights the history of eugenics and reasons we have to be cautionary about gene selection technology, but also examines some of the more promising uses of gene therapy. Goering observes that we need a decision-making process to help delineate what traits are rightful candidates for gene transfer, and that the medical model's distinction between treatment and enhancement, and the principle of beneficence, are less than illuminating in this regard. The medical model identifies the relief of pain and suffering, and premature death as the proper basis for human gene transfer research, but these are concepts that, like "disease" and "malady," are subjective and prone to social bias.

This problem is explored in specific detail in the film *Gattaca*. Vincent is portrayed as superior to the genetically enhanced elite, despite his "in-valid" birth status and a weak heart that threatens premature death at age 30.2. The film offers a number of character foils that call into question the links between "fitness," physical genetic enhancement, and the less tangible elements of personal character. While the genetically enhanced are physically and intellectually sound, they are deficient in willpower, impulse control, and grit. The film skillfully reveals the difficulty in pinpointing what counts as "fitness," and poses questions about how gene therapy might be used to cure and heal, rather than subjectively enhance future generations of people.

The most helpful aspect of Goering's essay is the procedure she offers for distinguishing between morally permissible and impermissible uses of human gene transfer, a procedure she adapts from the famous thought experiment of the veil of ignorance devised by John Rawls (Rawls, 1971). Originally used by Rawls to determine distributive justice for social goods and powers, the veil of ignorance refers to a rational act whereby individuals pretend not to know their own social interests and personal characteristics in order to determine objective preferences. Goering employs the veil of ignorance to resolve what traits we fairly and rationally desire for our children, and what traits we desire for them because of individual or social biases. She states:

> The veil of ignorance is a way to conceal from us the particular biases that our society has for the traits that are otherwise not genuinely physically desirable. When we put on this veil of ignorance, we assume that we do not know which society we will be living in—we do not know the physical or social details about the majority class, for instance. We then try to determine what physical traits would lead to clear advantages and disadvantages in any society. This test allows us to decide for our children and future generations what sorts of traits should not be genetically manipulated. (Goering, 2000, 336)

Goering uses this procedure to analyze the moral permissibility of gene therapy in a variety of cases, including race, sexual preference, beauty, Tay–Sachs disease, deafness, height, and intelligence. She finds that while the first three cases are eliminated as permissible candidates for gene therapy because they are not disadvantages in all possible societies, Tay–Sachs disease is a legitimate candidate because it is of universal disadvantage.

In Goering's analysis, the traits of deafness, height, and intelligence are more complicated. While deafness is a disadvantage in a society of mostly hearing people, it is not a liability when the majority of people are deaf. However, the hearing person does not seem disadvantaged in either society. There is an asymmetrical disadvantage in the case of deafness, then, because hearing people are not disadvantaged in a deaf society, but deaf people are disadvantaged in a hearing society. Alternatively, in the case of height, Goering finds the disadvantage to be symmetrical, because it is both a disadvantage to be short in a predominantly tall society, and inconvenient to be tall in a predominantly shorter society. Thus, height preferences are arbitrary and subjective, and *not* suitable candidates for gene therapy. The question of intelligence is even more complex. It would seem to always be an advantage, or at any rate never a (serious) disadvantage, to be smarter than those around you. Intelligence might then be an appropriate trait to select for. In short, Goering suggests that traits that are legitimate candidates for negative gene therapy are those traits rational people could agree would bring universal disadvantages (Tay–Sachs disease), and (maybe) asymmetrical disadvantages (deafness, blindness). Traits that we could seek to positively select for should be those that are not harmful in any possible society, and advantageous in some or all societies (e.g., intelligence).

Unlike Rawls, Goering is commendable for recognizing the limits of imaginative reasoning. Goering points out that we often deceive ourselves about what it is like to be another person, especially a person with a trait that strikes us as foreign, frightening, or repulsive. To correct for this limitation, Goering stipulates that ideal decision-making conditions about gene therapy should include first-person testimonies from those with the condition or trait being considered for therapy, and she divides the veil of ignorance into two steps. Step one involves bringing together differently abled individuals who discuss openly the benefits and harms, delights and difficulties, of living with various physical conditions. Step two consists of the deliberative representatives performing the thought experiment of the veil-of-ignorance strategy with the primary goal of picking out traits that should *not* be permissible candidates for genetic engineering, and the secondary, more difficult, goal of achieving consensus about what traits should be allowed as candidates for positive gene selection.

With these guidelines in place, Goering's procedural approach can be applied to many different genetic conditions and adapted to a number of pedagogical purposes. The procedure makes for a good in-class exercise, where students are organized into small groups tasked with the responsibility of coming to consensus about whether a certain condition should be permitted or banned as a candidate for gene therapy, and in what ranking (some examples are Down syndrome, Alzheimer's disease, Parkinson's disease, sexual preference, and gender). It can be a discussion topic coupled with *Gattaca*, where students are asked to identify genetically enhanced traits portrayed in the film and to consider whether they would be counted as morally legitimate by Goering's standards (some mentioned are myopia, manic depression, obesity, premature balding, and alcoholism). This in-class exercise can be extended over several periods by adding a research assignment where students research conditions to determine genetic linkages, and re-create step one by gathering firsthand testimonies of what it is like to have the condition under inquiry. Guest speakers, or biographical literature, can be used to enhance the offering of first-person testimonies.[1] The exercise also can become a service learning opportunity where students interview individuals with various conditions such as diabetes, Alzheimer's disease, etc., and then present their findings and conclusions to the class. Of course, Goering's procedural device can always be used as a paper topic or essay exam question.

Applied to *Gattaca*, Goering's procedure yields some interesting results. While it would appear initially that Vincent's weak heart is a disadvantage in any possible society because it causes premature death, it could be argued that in a society where life expectancy is only thirty, it would not be a disadvantage. This is not too far-fetched in terms of a normal conception of human existence, given that in some current societies life expectancies are not much higher. In a society where average life expectancy was thirty, this condition would be no different from developing a heart condition at age seventy-five in the current U.S. society. Another interesting application from the film is the twelve-fingered pianist, a condition that would seem to be permissible under Goering's scheme because it not a foreseeable disadvantage in any society, and is an advantage in societies populated with ten-fingered people.

However, despite the usefulness of Goering's test, it has some shortcomings. Goering acknowledges limitations with her procedure, but believes that that they do not belie its usefulness as a tool of elimination. She cites three critiques commonly leveled at Rawls: that the veil of ignorance only works within particular kind of liberal society; that it presumes rational creatures can successfully abstract away from their individual realities; and that it is not sufficiently attentive to what exists in this society. She responds by noting the need to assume some general conception of the human form to avoid bizarre but possible genetic traits (such as having gills) that are advantageous in some possible society and not harmful in our own. However, this seems itself to beg the question of how we can objectively determine what counts as an advantageous trait in some possible future without appealing to current physical and technological limitations and ungrounded social preferences.

In using this two-step veil of ignorance as a decision-making tool in my classroom, I have experienced two problems not fully explored by Goering that are worth highlighting. The first is that people tend to disagree about what constitutes an advantage and disadvantage, and they weigh benefits and burdens differently. For example, Goering considers that there might be disadvantages for a hearing person in a deaf society, such as being easily distracted, or experiencing schizophrenia because one is responding to stimuli not perceived by others, but she dismisses these as true disadvantages because they are not borne out by early childhood stories of hearing children raised in deaf families. However, these stories seem heavily influenced by other background conditions that need not apply in other scenarios, such as whether sounds available for perception are meaningful, and whether sound is accepted as a legitimate and real phenomenon. In a society where the wide majority of members are deaf and always have been, the existence of sound itself could be suspect, treated as akin to supernatural beliefs, or diagnosed as insanity. Furthermore, like the Taoist concept of yin/yang, what initially appears as an advantage might in other ways be a disadvantage, and vice versa—depending on variables too numerous to posit.

Goering's example of height also reveals this first problem. She assumes that height is a symmetrical disadvantage, because it is a disadvantage to be a giant among dwarves, or a dwarf among giants. However, not all disadvantages and advantages are static or equal. While taller people in a short society might bump their heads and not fit in standard furniture, they can reach tall items and carry heavy things. While people of shorter stature cannot reach tall items, they have the advantage of requiring fewer resources in terms of clothes, food, and housing, and might be better at building ladders, hiding, climbing, or tedious fingerwork. Although a tall person might be more feared in a short society, a short person in a tall society is easily crushed, robbed, or killed—a vulnerability, however, that might evoke sympathy and protectiveness. Deciding what is better or worse in the case of height, then, depends on many variables.

Because there are so many variations of possible societies and situations, it may seem better to limit our analysis to actual societies. But actual societies are marked by common injustices that skew our results in counterintuitive ways (such as the almost universal disadvantage of being born female that might render it a legitimate trait for gene therapy according to Goering's criteria). Therefore, I recommend that students consider a combination of both possible and actual societies in this thought experiment, and rank the most realistic harms and advantages after hearing first-person testimonies of *all* of those most affected by a condition.

The second problem with Goering's procedure has to do with the challenges of gathering first-person testimonies, not all of which are readily available or reliable. It may be impossible to gather meaningful first-person testimonies from people suffering from certain conditions such as mental illness, autism, or rare diseases. Nor can we gather first-person information from people with not-yet-existing conditions, and so we must speculate about what it might be like to live with these future possible traits. For example, in 2000, Eduardo Kac successfully fused a synthetic green fluorescent protein isolated from the genes of a jellyfish into the cells of a white rabbit, dubbing the act "transgenic art" (Kac, 2005). If we wish to determine whether the ability to fluoresce is a trait that we should desire for our children, at this point in time there is no one to offer a first-person narrative about what such a life is like. Furthermore, personal testimonies often conflict in reasonable ways over which traits are desirable in children. Even a trait as bizarre as glowing in the dark could be reasonably desired in some circumstances for its pragmatic usefulness or exotic rarity.

Yet another problem with the requirement of actual discourse is that we may have moral objections about gathering first-person testimonies about genetic traits that seem evil. Most of us do not wish to interview murderers or pedophiles about the joys and frustrations of their life experiences, even though these conditions may have genetic links and some of these individuals claim wrongful social discrimination, as do, for example, the pedophiles of NAMBLA, the North American Man–Boy Love Association. Finally, targeting groups with legitimate claims of social discrimination could only worsen their plight by marking their genes as questionably desirable. Although Goering is certainly right to require first-person testimonies as often and broadly as possible, this stage will sometimes have to be supplemented by rational and social imagination.

The murky intersections between genetics and moral character reveal a fascinating and thorny potential use of gene therapy raised by only marginally by *Gattaca* and Goering. In *Gattaca*, the promise of using gene transfer to affect moral character and social amiability is depicted as flawed. When someone crudely murders the director of the agency Gattaca, it is assumed that killer is the "in-valid" whose DNA is detected amon them, when in fact it is a member of the elite who was genetically programmed for nonviolence. Almost all of the genetically enhanced characters in the film exhibit moral defects, and the audience roots for Vincent even though he is a lawbreaker. The film suggests both that there is no predictable link between genes and moral character, and also that good and bad are subjectively ambiguous. While these points are valid, they do not form a complete picture. As James D. Watson notes, studies suggest that genetics play at least a partial role in the manifestation of conditions such as schizophrenia, manic depression, substance abuse, and some acts of violence (Watson, 1997, 632). While still highly speculative, it may be that dispositions toward pedophilia, sadism, suicide, or sociopathy have genetic markers (Katz and Marsh, 1985). Few of us would deny that some human behaviors are heinous, and that in considering genetic markers related to these behaviors the question shifts to include what is disadvantageous to society as well as the individual.

Thus, for more advanced students of bioethics, Goering's procedure can be used to decide whether gene therapy should be considered as a justifiable weapon against vice and mental illness.[2] For reasons I've explained, we may be tempted to skip the first phase of gathering first-person testimonials from people we consider grievously insane or immoral, but this step is no less important in these cases because judgments about such persons are persistent sites of social prejudice. We must then inquire whether presumed conditions of insanity and immoral behaviors are disadvantageous in every society. While schizophrenia and sociopathy would be a disadvantage in every possible society (since a society of schizophrenics or sociopaths would be highly disorderly for all), many other forms of mental illness and moral deviance are symmetrical, meaning that it is questionable to consider them as candidates for gene therapy according to Goering.

Pedophilia is another troubling case in point. The origins of pedophilia are still not clearly understood in that some research suggests that it is influenced by environmental factors (especially early sexual abuse), but other findings indicate a possible genetic connection.[3] Although it is a disadvantage to be a pedophile in a society where sex between adults and children is reviled and legally punishable, it would not seem to be a disadvantage to be a pedophile in a society more accepting of such practices (as was ancient Greece, for example). However, the case of pedophilia demonstrates the need to ask about the disadvantages of a trait not only for the individuals exhibiting the trait, but also for others affected by it. The primary moral objection to pedophilia is that it causes wrongful damage to the child victims of sex crimes perpetuated by adults, not that it harms pedophiles themselves. Working from the widely held presumption that children are not physically or psychologically equipped for sexual relations, especially sexual relations of inherently unequal power, the disadvantages of pedophilia to children could be claimed to be universal even in societies that accept such practices. This line of reasoning suggests that genetic intervention to suppress pedophilia would be justified, if feasible. But the truth of the conclusion depends upon whether children are universally (or predominantly) harmed by sexual relations with adults, or would be harmed even in societies where child–adult sexual relations are a respectable norm. Some advocates of pedophilia deny this premise, asserting that sexual relations with adults can be initiated by and pleasurable to children, that age is relative, and that context is important. Goering's test would suggest that we need to gather testimony from child victims who might verify this point of view. If verified, pedophilia would not be a universal disadvantage, and hence would not be a candidate for negative gene therapy, or gene repression. Furthermore, if it was possible that in some societies a disposition toward pedophilia would be an advantage, such a disposition might be a legitimate candidate for positive selection. (A similar case could be made for the disadvantages and advantages of a violent disposition.)

In considering this line of reasoning from the point of view of a prospective and actual parent, however, I find these results counterintuitive. Although I acknowledge that there may be some possible society where children would predominantly report that they are not damaged by sexual relations with adults (the polygamous community of Colorado City, Arizona, comes to mind), that being prone to violence offers some advantage in a good many possible and actual societies (as Machiavelli, Hobbes, and Nietzsche famously contended), and that various mental illnesses are sometimes wrongfully stigmatized, I would wish to prevent these traits and dispositions in my child if I could. I also appreciate, however, that many parents of children with these very traits feel more conflicted. This suggests the need to supplement Goering's use of Rawlsian liberalism with other moral frameworks. I will conclude by briefly explaining how the Ethics of Care and Libertarianism can be used to help dialectically evaluate the results we get from Goering's Rawlsian procedure.

The Ethics of Care is a moral framework attributed to the work of Carol Gilligan and Nel Noddings, who originally described it as a form of moral reasoning rooted in the traditional caregiving activities of women (Gilligan, 1982; Noddings, 1984). It is characterized by an emphasis on caring relations within families and other communities, contextual thinking about real people and situations, respect for the role of emotion and intimacy in moral reasoning, an understanding of the moral subject as relationally embedded, and a goal of maintaining balanced relationship. One of the features distinguishing the Ethics of Care from liberal moral frameworks is its relational emphasis. As mentioned earlier, Goering is right to suggest that we ought to begin an ethical analysis of gene therapy by gathering testimonials from people who live with conditions being considered for treatment, but this approach seems overly individualistic. Shouldn't we also gather testimonials from parents, siblings, caregivers, and others who are seriously affected by the traits in question? It is clear that we cannot thoroughly determine the pros and cons of living with a trait until we talk to the network of individuals whose lives are impacted by it. Such reports are essential for evaluating benefits and burdens to units, such as families, health facilities, and communities. In many cases we will find that the disadvantages of certain conditions can be assuaged by providing more community support, education, and understanding, and that genetic burdens and benefits are shared. This latter point is somewhat obscured in *Gattaca* because the main characters (even wheelchair user Jerome) are generally depicted as alienated and independent.

The Ethics of Care, unlike Goering's method, favors the use of actual scenarios over the hypothetical thought experiment of "possible worlds." However, the Ethics of Care is able to highlight the significance of counterfactuals (questions of "what if") within real-life scenarios. Expanding the scope of whose voice gets to inform the debate about the advantages and disadvantages of certain genetic traits reveals that the analysis of counterfactuals depends upon relational contingencies. For example, if we ask parents of children with traits that may someday be candidates for negative gene therapy whether they would have preferred that their children not have the traits in question, we are likely to get mixed responses. Parents-to-be, or parents of children with potentially fatal diseases, painful conditions, or disabilities acquired after birth might more quickly answer in the affirmative, while parents of children born with physical or emotional disabilities might exhibit more hesitation. However, it is important to balance this testimony with the likelihood that most of these persons, had they been interviewed prior to the birth of their child, would have preferred that this same child not have this condition, and would have changed the trait if they could have without causing any other deleterious affects on the child. This testimony must also be balanced against the counterfactual interests of others in society. For example, in the case of destructive or violent behaviors, it is important to weigh imaginative counterfactual testimonies of those who might be potentially harmed by these behaviors in considering the merits of gene therapy, including the subject. (The same is true for conditions that might be helpful.) The Ethics of Care can do a better job of tracking the interconnected interests of gene therapy because it construes the subject as necessarily embedded in nested relations that are by nature contextual.

Similarly, Libertarianism can also be brought to dialectically expand on Goering's suggestions. Goering concludes that gene therapy is not an evil deserving of prohibition, but that we should not, as libertarian Robert Nozick contends, allow market forces to decide what traits will be available for gene transfer or who gains access to it (Nozick, 1974). This certainly seems sensible given the bizarre, shallow, and harmful ways that such technology could be used by unscrupulous persons. There are also legitimate concerns that the Libertarian "personal service" model of delivering gene therapy will exacerbate the already widening gap between rich and poor—a problem illustrated in *Gattaca* by the improved standard of living afforded Vincent when he is an astronaut as opposed to a custodian (Buchanan, Brock, and Daniels, 2000).

However, most thinkers will not find the Libertarian position so easily defeated. Goering's procedural analysis suggests not only that should some diseases such as Tay–Sachs, cystic fibrosis, and cancer be privileged candidates for gene therapy, but that other traits such as height, skin color, sexual orientation, gender, etc. should be prohibited as candidates for gene transfer. As much as we might agree that fatal and painful conditions should be first in line for gene transfer cures, and that it would be shallow and prejudiced for people to prefer traits based on racist and sexist beliefs, liberty lovers will resist being told by governments, medical boards, or ethics committees what traits they can and cannot pass on to their children. Like Watson, who cautions that we should keep governments wholly out of genetic decisions, Libertarians appeal to the principle of autonomy to argue that individuals ought to retain economic and rational decision making rights over the use of gene therapies. Against the concern that the free market will allow for indiscriminate use of gene therapy that will result in unfortunate and unforeseen consequences, Libertarians might optimistically predict that genetic preferences will even out in much the same way they do through more conventional avenues, meaning that some people will select against certain traits and others will select in their favor. They could argue that infringement to the right of reproductive privacy should be reserved only in cases where substantial social harms can be proven as imminent.

Thus, while the Ethics of Care suggest that it does not seem advisable to go as far as Watson in demanding that national and international government have *no* role in regulating genetic treatments, Libertarianism holds that personal autonomy in the genetic marketplace should not be preemptively sacrificed out of fear of past eugenic abuses or the visage of an alien future. Synthesizing these two views reveals the need to evaluate the harms and potentials of gene therapy on a case-by-case basis that takes as its inspiration for "possible worlds" the historical and actual worlds of national and international communities. In the previously mentioned case of gender selection, an approach that synthesizes Goering's test with Care Ethics and Libertarianism reveals that the speculation that genetic preferences for boys and girls will "even out" is not true in all cases, and could yield unpredictable results. Although most of the gender selections that have been performed in the United States have been roughly equivalent for boys and girls, there are noted patterns of preference based on nationality, with the Chinese preferring boys and Canadians preferring girls. The subsequent gender imbalances that could result from such practices warrant our attention because although the ratio of boys to girls may even out in terms of global numbers, the disparities between the birth ratio of boys to girls in some countries (like China) are sufficiently problematic because of being rooted in ideologies of male superiority to warrant the United States to respect or impose a ban against gender selection. This case also shows the ambiguity of Goering's test without the addition of further context: It is true that it is a disadvantage to be born a girl in many countries, but it is also true that the practice of using gene therapy to select for girls could be of social benefit especially in societies that have a greater preference for boys. As a result, the United States may have a moral duty to refuse foreigners access to gender selection technology when it threatens to become a tool for perpetuating male domination, but may permit it when no such bias seems present, or when doing so promises to support relationships of equality (such as in allowing Chinese couples to select for girls). The universal individual "right" to select the gender of one's children needs to be balanced against the ways in which an international market perpetuates social inequities that violate the ideals of liberty and equality, but also against the ways that such technologies might be used to enhance these same ideals.

In conclusion, Goering's procedural adaptation of the Rawls veil of ignorance makes for a thought-evoking partner with the film *Gattaca* and other works of science fiction. Goering is correct that gene therapy is not an evil deserving of outright prohibition, but I have shown that the mere indication of a trait being a possible advantage or disadvantage in some society is not enough, absent contextual details and relational/liberal ideals, to indicate its permissibility or impermissibility as a candidate for gene therapy. In applying Goering's test to the difficult cases of deafness, height, mental illness, pedophilia, and gender, I explained why this test is subject to a number of variables that makes consensus difficult, and how it can be made more manageable with an unlikely synthesis of Care Ethics and Libertarianism that allows for a balance between relational and individualistic interests. Although they are certainly not the only decision making tools we will want to employ in drafting normative policy for gene therapy, both *Gattaca* and Goering are good starting points for assessing the ethics of calculated genetic intervention.

Notes

1. Two good sources of first person narratives, the first exploring deafness, and the second Huntington's chorea, are "A World of their Own" by Liz Mundy, *The Washington Post Magazine*, March 31, 2002, p. 22; and Alice Wexler, *Mapping Fate* (Berkeley: University of California Press, 1995). The Internet is also an increasingly strong resource for first-person testimonies, especially blogs posted by people living with terminal illnesses like cystic fibrosis and other genetic conditions.

2. Another interesting work of fiction exploring the connection between genes and criminal behavior is the novel *A Philosophical Investigation* by Philip Ker (New York: Plume Books, 1995). This futuristic novel describes how the British government uses genetic tests to identify citizens with criminal tendencies, gives them code names of philosophers to protect their privacy, and encourages them to undergo gene therapy. One person on the list, code named Wittgenstein, falls through the cracks and begins a serial killing spree.

3. Dr. John Money, the physician who became notorious for performing sex-reassignment operations on children based on his view that gender was the product of environmental socialization, championed the distinction between "affectionate" and "sadistic" pedophilia and upheld the possibility that some form of pedophilia has a genetic link. For a refutation of his views, see "Seduction or Genes" by Carl Pearlston, (http://www.ldolphin.org/genes.html).

References

Buchanan, A., D. Brock, and D. Wikler. (2000). *From Chance to Choice: Genetics and Justice*. New York: Cambridge University Press.

Gattaca. (1997). Columbia Pictures.

Gilligan, Carol. (1982). *In a Different Voice*. Cambridge, Mass.: Harvard University Press.

Goering, Sara. (2000). "Gene Therapies and the Pursuit of a Better Human." *Cambridge Quarterly of Healthcare Ethics*, vol. 9, no. 3 (summer), 330–341.
Huxley, Aldous. (1932). *Brave New World.* New York: Harper Collins.
Johnson, Carla K. (2006). "Wealthy Foreign Couples Travel to U.S. to Choose Baby's Sex," *Associated Press*, June 15.
Kac, Eduardo. (2005). "GFP Bunny," in *Telepresence and Bio Art—Networking Humans, Rabbits, and Robots*, Chapter 14. Ann Arbor, Mich.: University of Michigan Press.
Katz, Janet, and F. Marsh, eds. (1985). *Biology, Crime, and Ethics: A Study of Biological Explanations for Criminal Behavior.* Cincinnati: Anderson Publishing.
Ker, Philip. (1995). *A Philosophical Investigation.* New York: Plume Books.
Mundy, Liz. (2002). "A World of Their Own," *The Washington Post Magazine*, March 31, p. 22.
Noddings, Nel. (1984). *Caring.* Berkeley: University of California Press.
Nozick, Robert. (1974). *Anarchy, State, and Utopia.* New York: Basic Books.
Pearlston, Carl (2002). "Seduction or Genes" (http://www.ldolphin.org/genes.html).
Rawls, John. (1971). *A Theory of Justice.* Cambridge, Mass.: Harvard University Press.
Schiel, Betty A. (1992). "Ethics of Caring and the Institutional Ethics Committee," in *Feminist Perspectives in Medical Ethics*, eds. Helen Holmes and Laura Purdy, pp. 113–126. Bloomington: Indiana University Press.
Sherwin, Susan. (1996). "Feminism and Bioethics," in *Feminism and Bioethics*, ed. Susan Wolf, pp. 44–67. New York: Oxford University Press.
Watson, James D. (1997). "Genes and Politics," in *Journal of Molecular Medicine* (Sept.), pp. 632–636.
Wexler, Alice. (1995). *Mapping Fate.* Berkeley: University of California Press.

Best Interests and *My Sister's Keeper*

Terrance McConnell
UNIVERSITY OF NORTH CAROLINA–GREENSBORO

For most of medicine's history, the power to make decisions was vested in physicians. It was assumed that in the context of medicine physicians knew what was best for patients and would do what was best for them. Such a view is typically referred to as medical paternalism (or the Hippocratic view). In the last half century (or so), however, this view of the role of doctors in decision making has been gradually dismantled.

Patients first won the right to refuse recommended medical treatment. In a case in New York State, in 1914, Justice Cardoza ruled that "every human being of adult years and sound mind has a right to determine what shall be done with his own body."[1] There were a number of cases involving competent Jehovah's Witnesses who, for religious reasons, refused blood transfusions. When doctors administered such treatment in order to save the patients' lives, courts ruled that the patients' rights had been violated. Forcing medical interventions on competent patients violates their right of bodily integrity.[2] In medicine, no means "no."

Until the second half of the twentieth century, yes meant "yes." If patients agreed to a doctor's treatment recommendation and the doctor was in no way negligent, then any untoward results were not something for which the physician was legally liable. But all of that changed with the evolution of the right to informed consent. In a series of legal cases from 1957 to 1972,[3] the courts established that a patient's agreement did not fully authorize treatment unless it was informed. In order for a patient's consent to be informed, minimally that patient must be told about the expected benefits of the treatment, the risks, the alternatives (if any), and the likely consequences of refusing the treatment.

Affirmation of the rights to refuse treatment and to informed consent together constituted an assault on the paternalistic model of the health care provider–patient relationship. In the old paradigm, physicians made all decisions; under the new paradigm, the relationship is more collaborative. While it is certainly an exaggeration to say that patients have taken control of the relationship, they at least have attained a greater degree of autonomy. But who should make decisions for patients who are not competent?[4]

Until the second half of the twentieth century, physicians made decisions for patients who were not competent. But with the evolution of patients' rights and the recognition that patients sometimes do not want treatment that doctors recommend, it was natural to put the question of surrogate decision making on the table. In several highly publicized cases—including the case of Karen Quinlan and the case of Nancy Cruzan—families openly challenged the appropriateness of treatments that doctors were administering to patients who were no longer competent. The typical structure of these cases (in the 1970s and 1980s) was this: The health care team wanted to continue life-sustaining treatment for an incompetent patient, while the surrogates wanted to stop the intervention because the patient's prognosis was poor.[5] The questions underlying these disputes were these: Who is the proper decision maker for an incompetent patient? What principles should guide surrogate decision makers? A consensus has emerged about the proper answer to the second question.

In making medical decisions for incompetent patients, surrogates should be guided by three principles, arranged hierarchically.[6] According to the advance directive principle, if the patient has executed either a living will or a health care power of attorney, then surrogates should make medical decisions in accordance with the document in question. The value underlying this principle is autonomy; patients who have executed a living will or health care power of attorney have made a deliberate decision, while competent, about their treatment preferences in end-of-life situations. If the patient has no advance directive, then surrogates should be guided by the substituted judgment principle. According to this precept, surrogates should make medical decisions for incompetent patients on the basis of what the patients themselves would have wanted. Evidence of what a patient would now want might include past conversations, statements, letters, comments on publicly discussed cases, and the like. Surrogates should substitute the patient's judgments for their own. This principle too promotes autonomy. If the patient has no advance directive and there is no evidence of what the patient would have wanted in the situation in question, then surrogates should make decisions based on the best interest principle. Here the underlying value is beneficence; autonomy is not relevant.

There are two situations where it seems that the best interest principle is the only viable option. One is when an incompetent adult patient was never previously competent.[7] The other case concerns children—individuals who are *not yet* competent. It is widely held that parents should follow the best interest principle when they make decisions for their children. Determining what is in the best interests of children is complicated by the fact that they will (normally) become individuals who have the capacity to make their own decisions; in part, what is best for children is what will enable them to develop their own decision making capacities.

As many have observed, however, society does not literally hold parents to the best interest principle.[8] If parents have more than one child, what is best for one may be in conflict with what is best for another. And parents have legitimate interests of their own, some of which may conflict with what is best for their children. As Allen Buchanan and Dan Brock put it, "The case for legitimate departures from the best interest principle as a guidance principle in decision making for children depends exclusively upon the fact that optimizing for the sick child may conflict with legitimate interests of other *individuals* within the family."[9] And Lainie Friedman Ross tells us, "In liberal communities, the state tolerates a wide range of distributions among families provided that the parents provide their children with a threshold level of each primary good."[10] Ross's remark asserts, quite plausibly, that in cases of intrafamilial conflict, parents have much latitude, but not unlimited freedom, in the decisions they make for their children. But in these cases the devil is in the details. In hard cases we must ask, "When does society have a right to intervene when parents make a decision for one of their children?" This is a question with which we must wrestle in Jodi Picoult's *My Sister's Keeper*.[11]

This novel is about Sara and Brian Fitzgerald and their children, Jesse, Kate, and Anna. At age three, Kate was diagnosed with acute promyelocytic leukemia (APL). Her parents were told that the survival rate of APL was 20 to 30 percent, if treatment were begun immediately (p. 33). Chemotherapy was recommended for Kate. The hope was to achieve remission and perform an autologous harvest, retrieving Kate's own cells and reinstilling them. If that failed, she would be put on a national registry for a bone marrow donor (pp. 62–63). It did fail, and no family member was biologically suitable to be a donor for Kate. This meant that Kate would be placed on a national registry seeking an unrelated, matched donor. But the odds of finding such a donor were slim. Sara, however, apparently based on a suggestion from one of Kate's doctors, decided on another strategy (pp. 67, 72). She wanted to have another child, one she was sure would be a matched donor for Kate. Using in vitro fertilization (IVF) and preimplantation genetic diagnosis (PGD) to assure success, Brian and Sara had Anna.

Upon Anna's birth, the umbilical cord was carefully preserved and cord blood was extracted for Kate. That was the first of many contributions that Anna made to her sister. During the next thirteen years, she also donated lymphocytes (three different times), bone marrow, granulocytes, and peripheral blood stem cells (p. 21). As a result of the bone marrow transplant, Kate experienced graft-versus-host disease. Physicians gave her steroids and cyclosporine to control it, but that adversely affected her kidneys. So Kate, at age sixteen, needed a kidney transplant. And Anna, conceived to be a lifesaving resource for Kate, would be the perfect kidney donor. There was one small problem, however. Anna said "no." Readers assume that she has had enough; the full story behind her refusal is not clear until the end. Sara, the desperate mother, will not accept Anna's answer. So Anna seeks the services of Campbell Alexander, an attorney; Anna is seeking "legal emancipation for medical purposes" (p. 23).

The premise for Picoult's novel was undoubtedly derived from a real-life case.[12] In August 2000, Lisa and Jack Nash gave birth to a son, Adam. The Nashes' six-year-old daughter Molly had Fanconi anemia. She needed a bone marrow transplant, but no suitable donor was available. Using IVF, the Nashes created fifteen embryos. Of these fertilized eggs, two were both free of the disease and capable of being a matched donor for Molly. Only one of these two embryos was developing normally. That embryo was transferred to Lisa Nash; it implanted in her uterus, and became Adam. Editorial columnists across the nation wrote about this case, usually critically.[13] People wondered what impact it would have on a child if he knew that he was deliberately conceived to be a donor for a sibling. *My Sister's Keeper* explores this issue. Multiple positions are represented. Readers cannot help but ask, "Are the Fitzgeralds good parents?" Brian and Sara asked themselves the same question several times (e.g., p. 353). Throughout the novel, many things go wrong for Jesse and Anna. Each seems unhappy on many occasions. Sara and Brian's relationship deteriorates (e.g., pp. 197, 235, and 259). But to what extent, if at all, are these troubles due to the unique circumstances under which Anna was conceived?

Some of the difficulties experienced by Jesse and by Anna are ones common to most children. For example, as a young child, Jesse tormented Kate (p. 29). And Jesse would sometimes protest that his parents were giving Kate preferential treatment (p. 68). These are small matters that nearly all families experience. But some of Jesse's troubles are much deeper. He describes himself as "the Lost Cause" (p. 15), and frequently exhibits a lack of self-worth (p. 98). He thinks that he should be dead (p. 94). We then learn that it is Jesse who is the local pyromaniac wreaking havoc in the community (p. 95). This is especially significant since Brian is a captain of the local fire-fighting squad. Both Kate (p. 161) and Jesse himself (p. 247) recognize that much of his antisocial behavior is designed to get the attention of his parents. Jesse even recognizes that his parents sometimes try to compensate for having ignored him. He tells readers that often after one of "Kate's episodes" his parents will present him with a "guilt gift" (p. 244). Of course, when he perceives his parents' motives in this way, it merely exacerbates the situation. These latter problems are ones that are apt to plague any family that has a child with a serious chronic illness. Jesse himself gives voice to this when at one point he says to his mother, "You don't know what it is like being the kid whose sister is dying of cancer" (p. 267).

Anna is obviously a more interesting case since it was she who was created to be a donor for Kate. Some of Anna's troubles plague many teenagers. She complains about not being noticed (p. 22, p. 40), she wishes she were someone else (pp. 137–138), and she professes self-hatred (p. 339). And like Jesse, Anna experiences difficulties because there is a chronically ill child in the family. For example, she complains that her mother's world revolves around Kate (p. 11). But some of what Anna goes through seems directly related to the circumstances of and reasons for her creation. She sometimes thinks of herself as "a collection of recessive genes" (p. 49), and she is dubbed by some, "Rhode Island's first designer baby" (p. 182). Anna was an avid hockey player (virtually unbeknownst to her parents), and one summer she was accepted to participate in the prestigious "Girls in Goal Summer Hockey Camp" in Minnesota. Her excitement was dashed, however, when her mother told her that she could not attend this two-week session because she might be needed if something happened to Kate (pp. 268–269). Readers also learn that Anna often thinks of her identity as being nothing but a donor for Kate (p. 138), and she memorably tells her mother that "Anna" is a "four-letter word for vessel" (p. 251). It is safe to say, then, that Anna experienced some special difficulties because she was created to be a donor for Kate.

My Sister's Keeper raises at least two important ethical issues. First, should people have the freedom to use IVF and PGD to create a child to be a donor for a sibling? Second, if so, at what point have parents demanded too much of the donor? When and for what reasons may society intervene?

Concerning the first question, I take it as a given that if society restricts people's options it must have a good reason for doing so. The least controversial reason for prohibiting conduct is to prevent harm to nonconsenting parties. Who might be harmed if parents are allowed to use biotechnology to create a donor sibling for a sick child? The only apparent candidate is the child so created. Editorial columnists who criticized the Nashes opined that children created to be donors would have serious psychological problems. If Picoult's novel is interpreted to provide an answer to this first question, it does not seem to counsel proscribing the option in question, at least not based on the harm that it does to the child created. Anna Fitzgerald was created to be a donor for her sister Kate. But in no way are readers led to believe that Anna's life is horrible. In many ways, she seems to be a normal child experiencing the same joys and frustrations that most other kids do. Moreover, of the problems Anna does experience, most are unrelated to the reason that she was created. In many cases, Anna's angst is like that of most other teenagers; in other cases, the troubles are ones that any family with a seriously ill child could have. Finally, we would seem to lapse into philosophical paradox if we were to say that the very act of creating Anna harmed her.[14] It is mind-boggling, at best, to say that nonexistence is better than existence, for there is nothing against which to compare the latter.

But what about the second question? Did the Fitzgeralds demand too much of Anna? Was society justified in intervening in this case? The verdict here is less clear. Once Anna was created, she was a person with rights, and those rights put obligations on others, including her parents. As noted earlier, however, society does not literally demand that parents make decisions that are always in their child's best interest. Trade-offs are permitted. It is only extreme cases that are obvious. Parents are not permitted to refuse lifesaving treatment for a child who can live a fulfilling life; in such a case, society will intervene to protect the child. At the other end of the spectrum, if Anna had been an only child and if attending the summer hockey camp were in her best interests, society would not have interfered had Brian and Sara refused to allow her to go. But the problem in *My Sister's Keeper* falls between these two extremes. A series of escalating demands is made on Anna. Taking cord blood, obviously, imposes no burden on Anna. The other interventions impose burdens of various degrees, the most serious of which is the bone marrow donation. But even that, though painful, imposes no serious risks on Anna. The ante is upped considerably, however, when Sara effectively tells Anna that she must contribute one of her kidneys to Kate.

Campbell Alexander represented Anna in her legal quest to gain medical emancipation from her parents. Sara Fitzgerald, herself an attorney, represented the family in opposing this petition. As the case was argued in court, two separate questions were sometimes merged. Were Sara and Brian's decisions in Anna's best interest? Is Anna mature enough to decide for herself? We can be sure that the second question would not even get a serious airing unless there were reason to believe that the answer to the first question was negative.

In addition to trying to demonstrate Anna's maturity, Campbell Alexander advanced two main arguments. First, he said that in the United States no person is required to assist another (p. 294, p. 305). In effect, so-called Good Samaritan acts are not legally mandatory. Anna's donating a kidney to Kate would be a Good Samaritan act, so she cannot be compelled to do it.[15] His second argument appealed to Anna's interests. In cross-examining one of Kate's physicians, Dr. Chance, Campbell established that the hospital itself acknowledged that Anna had been subjected to risks. Campbell had Dr. Chance read from Anna's consent form when she donated bone marrow to Kate. It said, in part: "These risks may include, but are not limited to: adverse drug reactions, sore throat, injury to teeth and dental work, damage to vocal cords, respiratory problems, minor pain and discomfort, loss of sensation, headaches, infection, allergic reaction, awareness during general anesthesia, jaundice, bleeding, nerve injury, blood clot, heart attack, brain damage, and even loss of bodily function or of life" (p. 335). Later, during the same cross-examination, Campbell handed Dr. Chance a flyer from the nephrology department of his own hospital. It contained information about being a living kidney donor. Campbell asked Dr. Chance to read a section concerning risks to the donor that he had highlighted. It said: "Increased chance of hypertension. Possible complications during pregnancy. Donors are advised to refrain from contact sports to eliminate the risk of harming their remaining kidney" (p. 336). Campbell also elicited from Dr. Chance the admission that doctors did not know the long-term effects of growth factor shots (which Anna had undergone twice prior to harvesting). The upshot of Campbell's second argument was that Anna had already been subjected to risks, that the most serious risks were yet to come, and none was for Anna's benefit.

Sara attempted to rebut Campbell's second argument through the testimony of a child psychiatrist. The psychiatrist claimed that a child is harmed when a sibling dies. And because Anna and Kate are very close, it is reasonable to think that Anna herself will benefit by making a lifesaving kidney donation to Kate (pp. 364–365). This recalls an actual case, *Strunk v. Strunk*, in which a family sought to harvest a kidney from a twenty-seven-year-old severely retarded man to give to his seriously ill brother. The court permitted the act, arguing that because of the special relationship between the donor and his brother, it was actually in the donor's best interest.[16] Sara was making a similar argument.

Sara countered Campbell's first argument by claiming, in effect, that the Good Samaritan analogy does not apply within the family. In her closing argument, she said, "At the beginning of this hearing, Mr. Alexander, you said that none of us is obligated to go into a fire and save someone else from a burning building. But that all changes if you're a parent and the person in that burning building is your child." She continues, "In my life, though, that building was on fire, one of my children was in it—and the only opportunity to save her was to send in my other child, because she was the only one who knew the way. Did I know I was taking a risk? Of course. Did I realize it meant maybe losing both of them? Yes" (p. 406). Earlier in his testimony, Brian had said something similar. "You don't know what it's like . . . until your child is dying. You find yourself saying things and doing things you don't want to do or say. . . . I didn't want to do that to Anna. But I couldn't lose Kate" (p. 344). The position of the Fitzgeralds seems to be this: If the life of one child is in danger, families should be free to put another child at some risk if doing so is the only hope of saving the first child. Sara and Brian knew that they were subjecting Anna to procedures and interventions that were not in her best interest. But they felt they should have the discretion to do this, at least to a point.

Campbell's closing statement acknowledges the difficulty of the situation facing Sara and Brian. "We know that the Fitzgeralds were asked to do the impossible—make informed health-care decisions for two of their children, who had opposing medical interests" (p. 407). This is one reason that society does not demand that parents always act in ways that always promotes the best interests of their children. Doing so is impossible; some trade-offs are permissible. But how far may parents go? In this case, Judge DeSalvo ruled in Anna's favor. He said, however, that Anna's age was not the most important factor. Rather, he reasoned, "[S]ome of the adults here seem to have forgotten the simplest childhood rule: You don't take something away from someone without asking permission" (p. 409). It is doubtful, however, that the guiding rule is quite that simple. For if we imagine Anna making her challenge at an earlier point—say, when they want to extract her bone marrow for Kate—it is not at all clear that a court would have ruled in her favor. Yet this would also be a case of taking something from her without permission. Whether society is willing to intervene with parental decisions in these kind of cases seems to depend on the degree of risk and sacrifice involved for the would-be donor.

Lainie Friedman Ross, in *Children, Families, and Health Care Decision Making*, argues against the best interest principle and in favor of what she calls "the model of constrained parental autonomy." On this view, parents are permitted to make decisions that fail to maximize the interests of one of their children in order better to promote the interest of another family member, provided that the basic interests of all children are met. But clearly this requires "drawing lines." And in a chapter entitled "The Child as Organ Donor,"[17] Ross argues that parents have limited discretion to decide that their children will be organ donors for another family member. Two variables are important in understanding the position that she defends: One is whether the child is appropriately considered competent; the other is the degree of risk involved in the donation. She distinguishes among "minimal risk," "a minor increase over minimal risk," and "more than a minor increase over minimal risk."[18] Given the realities of the procedures, Ross contends that bone marrow donation is plausibly viewed as a minor increase over minimal risk. In such a case, she thinks, parents should have the freedom to direct their child to make such a donation to another family member. A kidney donation, however, is more than a minor increase over minimal risk. In that situation, Ross argues, parents should not have the authority to allow such a donation if the child is *not* competent. If the child *is* competent, however, then Ross believes that the donation should be permitted if the child herself agrees. In short, when the child is competent and the donation procedure involves more than minor increase over minimal risk, then both the parents and the child must agree; if either dissents, the donation should not occur.

Assuming that Anna would be deemed competent, then Ross's position would entail that the Fitzgeralds should not be permitted to direct her to donate a kidney to Kate. *If* the distinction between a minor increase and a more than minor increase over minimal risk can be maintained, *if* there is a plausible link between subjecting someone to a more than minor increase over minimal risk and failing to protect that individual's basic interests, and *if* donating a kidney is appropriately regarded as more than a minor increase over minimal risk, then we may have a framework for supporting Judge DeSalvo's decision, though not for the reason that he gave. The framework may also lend credence to my speculation that had Anna's challenge occurred with one of the earlier procedures, she would have lost.

Sara Fitzgerald would undoubtedly agree with Ross that parents should be free to make trade-offs in pursuing the best interests of their children.

She is apt to argue, however, that when the stakes for her sick daughter are life and death, she should have the freedom to impose somewhat greater risks on her healthy child. But even Sara would concede that there are lines that may not be crossed. She would not agree with Ross, however, that kidney donation crosses the line.

However we view this particular case, determining how far below a child's "best interest" society should allow parental decisions to go is a difficult matter. And Jodi Picoult's novel is an excellent vehicle for exploring this complex issue.

Teaching My Sister's Keeper

I have used this novel in three different classes, and it has been well received. I do not assume that there is a best way to teach this book. I will describe briefly, however, the approach that I have adopted.

I have found it useful to begin by detailing Kate's medical condition *before* Anna was conceived. It helps to ask students what they would do were they in the situation of the Fitzgeralds. Typically this establishes at least some sympathy for the excruciating choices that this family must make.

My Sister's Keeper is a long novel, more than four hundred pages in length. This creates a pedagogical issue: Students are not likely to have read the entirety of the book by the time of the first class meeting at which it is scheduled. This is not a major obstacle, however. The novel details Kate's medical problems early and also provides an account of all that is demanded of Anna. This is the information that is needed in order to prompt students to discuss the key ethical issues. If my experience is a good guide, most students will have completed the novel by the second class meeting (on a schedule where classes meet twice a week).

My ultimate goal in using this book in an ethics and genetics course is to address two fundamental issues: What principles should parents use in making medical decisions for their children? And when is society justified in interfering with parental decisions in order to protect the interests and rights of a child? By discussing Kate's extraordinary problems and the Fitzgeralds' limited options, it is usually easy to get students to arrive at the best interests principle and to see its limitations. Getting to the second question may require a more circuitous route. Both Jesse and Anna have many problems throughout the novel, and I have found it useful to have students compile a list of these. I then ask them why these problems have arisen for Jesse and Anna. The goal is to get students to distinguish the three sources of difficulties mentioned in the main body of this essay. This provides an added bonus of being able to explore the challenges that any family who has a chronically ill child faces. The toll that Kate's illness takes on Sara and Brian's relationship illustrates this same point.

From here, one can move naturally to the key questions (related to ethics and genetics): Should parents be free to use PGD to create savior siblings? If not, why not? If so, how much may parents demand of these children? Details matter here. The sorts of risks that Anna incurs by contributing bone marrow to Kate are different from the risks of donating a kidney. But it is exactly these differences that show the difficulty of the issue: Just how far below the savior sibling's best interests do we allow parents to go in promoting the well-being of another child?

Notes

1. *Schloendorff v. N.Y. Hospital* (1914).

2. An exception to this is when a patient's refusal of treatment poses a risk of harm to nonconsenting parties. An example is when a patient refuses treatment for a contagious disease that can be spread through casual contact (such as tuberculosis). In this situation, the patient's right of bodily integrity has been overridden by a more important consideration.

3. See Ruth Faden and Tom Beauchamp, *A History and Theory of Informed Consent* (New York: Oxford University Press, 1986), Chapters 3 and 4.

4. I do not here define "competence." I take it to mean, roughly, the capacity to make one's own decisions. Competence is not always an all-or-nothing matter. Individuals may be able to make some decisions but not others. On these matters, see Allen Buchanan and Dan Brock, *Deciding for Others: The Ethics of Surrogate Decision Making* (New York: Cambridge University Press, 1989), Chapter 1.

5. In the most common type of case, the patient was in a persistent vegetative state.

6. The definitive work on this topic is Buchanan and Brock, *Deciding for Others*. See especially Chapter 2.

7. The most widely discussed case of this sort is that of Joseph Saikewicz. See Buchanan and Brock, *Deciding for Others*, pp. 114–115 and 124–126.

8. See Buchanan and Brock, *Deciding for Others*, pp. 235–237, and Lainie Friedman Ross, *Children, Families, and Health Care Decision Making* (New York: Oxford University Press, 1998), especially Chapters 1 and 3.

9. Buchanan and Brock, *Deciding for Others*, p. 236. Emphasis in original.

10. Ross, *Children, Families, and Health Care Decision Making*, p. 8.

11. Jodi Picoult, *My Sister's Keeper* (New York: Washington Square Press, 2004). Page references provided parenthetically in the text are to this work.

12. We need not speculate about this. In an interview with the author published

at the back of the paperback edition, Picoult indicates that a case in the United States that she heard about (she mentions no names) influenced this novel.

13. See, for example, Ellen Goodman, "When a Child Becomes Spare Parts," *News & Record* (Greensboro, N.C.), October 9, 2000.

14. The seminal work on this topic is Derek Parfit, *Reasons and Persons* (New York: Oxford University Press, 1984), especially Chapters 16 and 18.

15. There is a legal case specific to organ donation that is often cited to show this point. The case involved a person, Robert McFall, attempting to procure bone marrow from a cousin. The judge ruled against Mr. McFall. See *McFall v. Shimp* (Allegheny County Court [Pennsylvania], 1978).

16. *Strunk v. Strunk*, Court of Appeals of Kentucky (1969).

17. Ross, *Children, Families, and Health Care Decision Making*, Chapter 6.

18. The position described in this paragraph is developed in Ross, *Children, Families, and Health Care Decision Making*, pp. 112–120.

Genetic Discrimination in Health Insurance

An Ethical and Economic Analysis

Ben Eggleston
UNIVERSITY OF KANSAS

1. Hopes and Fears Aroused by Genetic Testing

In the 1997 movie *Gattaca*,[1] DNA is destiny. Practically instantaneous analyses of individuals' entire genomes reveal their bearers' precise propensities toward heart disease, depression, nearsightedness, and a host of other major and minor physical and mental conditions. In the opening minutes of the film, on-screen text sets the events in the "not-too-distant future," but few who saw the movie upon its release would have predicted the advances in genetic testing that would be made in the ensuing decade. To be sure, the world of *Gattaca* remains distant, and has grown at least a little more illusory: Currently we have genetic tests for only a small fraction of the diseases and other conditions that appear to be hereditary, and we know too much about the pervasiveness of gene–environment interactions, in the production of individuals' phenotypes, to countenance any strong form of genetic determinism for most significant ailments.[2] Nevertheless, correlations between genes and diseases are being discovered at an impressive rate, making genetic testing one of the fastest growing areas of health care.

Genetic testing has obvious benefits. For example, Ashkenazi Jewish women who test positive for a mutation in the *BRCA1* or *BRCA2* tumor-suppressor gene have a lifetime breast cancer incidence of 82 percent, compared to an incidence of less than 20 percent for the female population at large,[3] and although negative results offer no assurance of remaining cancer free (since most cases of breast cancer appear to have other causes), positive results can lead to more vigilant screening, more effective treatment, and the saving of lives. Similarly, a young man with a family history of Huntington's disease, and who has yet to experience the (normally mid-life) onset of the disease's debilitating symptoms, may feel that only a genetic test will enable him to make responsible decisions on such major questions as whether to have children and what kind of career to pursue. Admittedly, such foreknowledge, especially in the case of a disease that has no treatment or cure, is not always an unalloyed blessing, and some individuals choose ignorance rather than run the risk of being given what they would regard as a death sentence.[4] But for many people, the explosion in genetic testing is a source of nearly priceless information.

Equally, though, the explosion in genetic testing arouses fears of discrimination—in employment; in health, life, and disability insurance; and more amorphously in the creation of a "genetic underclass." Indeed, it is said that these fears are the greatest source of public concern about the ongoing revolution in genetics.[5] How widely genetic discrimination is being practiced, where it is legal, is unclear. As early as 1996, a study of the perceptions of members of genetic support groups found that 22 percent believed they had been victims of genetic discrimination in health insurance,[6] while a later study suggested that such fears were largely unfounded.[7] Despite disagreement about the extent of actually occurring genetic discrimination, a broad consensus has formed in opposition to such discrimination. Naturally, positions vary from "no amount of such discrimination is acceptable"[8] to more nuanced views, but the general movement against genetic discrimination is unmistakable,[9] and governments in Europe as well as North America have taken steps to prohibit various forms of it.[10]

Genetic discrimination is often treated as a single phenomenon, but it can arise in many different contexts, as noted earlier, and these different contexts present different issues. In this chapter, I focus on genetic discrimination in the context of health insurance, both because of the obvious importance of health care to quality of life and because of certain distinctive features of insurance as a product sold in a competitive market. The conventional view is that such discrimination is immoral and ought to be illegal. The prevalence of this view is understandable, given the widespread belief, which I endorse, that every individual in a society as affluent as ours has a basic right to affordable health care. But prohibiting genetic discrimination in health insurance is not an effective way to protect this right. On the contrary, I argue here that because of the nature of health insurance, such a prohibition is misguided, and that its worthy aims must, instead, be pursued though reforms in our country's system of publicly provided health care.

2. *Fairness in the Marketplace*

Much of the opposition to genetic discrimination in health insurance stems from the belief that such discrimination would be unfair. There are several arguments for this claim, and while several of the most prominent ones may initially seem intuitively plausible, careful scrutiny shows them to have significant shortcomings. Furthermore, a prima facie case can be offered in support of the claim that it would actually be unfair to prohibit health insurers from differentiating among prospective customers on the basis of their genes.

No doubt the most prominent argument for regarding genetic discrimination in health insurance as unfair is that people cannot control the genes they are born with, and it is unfair for people to be disadvantaged (in having to pay more for health insurance, for example) due to factors beyond their control. But these claims, though plausible, do not imply that it is unfair for insurers to set higher than average premiums for people whose genes indicate that they will probably have greater than average health care needs. To see this, bracket the idea of health insurance for a moment and imagine a small-town doctor, Dr. Smith, who has a few hundred patients who pay her directly rather than having health insurance. Some of her patients, due to factors beyond their control, need more health care than others. Since Dr. Smith charges her patients in accordance with the goods and services they consume, these unhealthy patients have to pay more, to maintain their well-being, than her other patients do. Thus, they are disadvantaged (in having to pay more to maintain their well-being) due to factors beyond their control. But is it unfair of Dr. Smith to charge her unhealthy patients more? This does not seem to follow. Instead, it seems more reasonable to locate the unfairness in the failure of the community at large to bear the cost of the extra health care that its unfortunate members need. Analogous reasoning can, and should, apply to the case of health insurance: It is not unfair for insurers to set higher prices for people they predict will require more health care; rather, it is unfair that society at large does not bear more of these extra, and undeserved, burdens. I take up this point again in section 4.

A second argument for the unfairness of genetic discrimination in health insurance challenges the analogy between health insurance and the case of Dr. Smith by claiming that insurance is different from a fee-for-service business such as the medical practice in which Dr. Smith is engaged. Specifically, it claims that insurance is inherently a form of risk sharing, in which the fates of the lucky and the unlucky are bound together and borne by all equally.[11] Admittedly, there is a grain of truth in this view of insurance, in the sense that no insurer could remain solvent if its lucky customers (those who do not end up needing much insurance) do not pay premiums sufficient to cover the extra claims of its unlucky customers (those who end up needing a lot of insurance).[12] But the idea of risk sharing does not show any unfairness in the setting of higher premiums for people with greater expected insurance needs. To see this, imagine that Dr. Smith offers her patients the following option: Instead of paying her on a fee-for-service basis, each patient can ask her to specify a monthly premium that they can pay instead. (Some patients might prefer the predictable monthly payment to paying on a fee-for-service basis, even if the premiums cost them more in the long run.) Dr. Smith, obviously, is now selling not only health care, as before, but also health insurance, with herself as the provider, to those patients who prefer that mode of paying for their health care. Is it unfair for her to set different premiums for different patients, depending on her predictions of their future health care needs? The idea of risk-sharing does not seem to give us reason to think so. Of course, in order for Dr. Smith's insurance option not to be a net loss for her, the premiums of her patients who end up needing less health care than she anticipates must be sufficient to cover the extra expenses of her patients who end up needing more health care than she anticipates. To this extent, Dr. Smith's patients who buy insurance are involved in risk sharing with one another. But the idea of risk sharing does not show that it is unfair for her to set different premiums for different people, depending on their predicted future health care needs.

A third argument purporting to establish the unfairness of genetic discrimination in health insurance invokes the fact that such discrimination typically involves setting higher premiums for customers because of the anticipated onset of conditions for which they are currently asymptomatic. Proponents of this argument claim that it is one thing to set higher premiums for customers with preexisting conditions (as is now standard practice), but that it is quite another to set higher premiums for customers with nonexisting, merely anticipated, conditions.[13] The comparison with preexisting conditions, however, does not undermine, but actually bolsters, the case for taking customers' genes into account in the setting of premiums. The rationale for setting higher premiums for customers with preexisting conditions is simply that they are likely to have greater than average health care needs in the future, and premiums should be proportional to anticipated needs. As a result, any source of information facilitating predictions of a prospective customer's future health care needs is as appropriate to consider as any other. If a customer is likely to have greater than average health care needs in the future, it is irrelevant whether this information is derived from genetic testing as opposed to a preexisting condition.

A final argument alleging the unfairness of genetic discrimination in health insurance is based on the fact that since genetic testing is still a relatively new field, and since the results of genetic tests are liable to be misunderstood by insurance underwriters, some customers' genetic predispositions to various diseases may be overestimated and their premiums set higher than they should be.[14] This argument, however, assumes that it is unfair for a prospective customer to be quoted a premium that has been inflated by an error on the part of the insurer. But is this assumption true? If an insurer overestimates the future health care needs of a prospective customer who smokes, or who has a preexisting condition, and sets a higher than appropriate premium as a result, we don't accuse the insurer of treating the customer unfairly; instead, we note that the insurer is opening itself up to losing some of its customers—all of the ones whom it is overcharging—to insurers that estimate future health care needs more accurately. The situation is analogous to one in which a mechanic overestimates the work that will be required to fix my car and quotes me a higher price than he would if he estimated the scope of the job accurately. It may be inconvenient for me that the mechanic quotes me a price based on a miscalculation—I may have to keep shopping around, or I may pay the higher price because I don't know any better—but it isn't a case of unfairness.

So there are significant shortcomings in the main arguments for the unfairness of genetic discrimination in health insurance. Moreover, a prima facie case can be offered in support of the claim that it would actually be unfair to prohibit insurers from differentiating among prospective customers on the basis of their genes. It is a basic tenet of free markets that actors in such markets, be they individuals or firms, are free to enter into those commercial transactions they believe to be advantageous, and to decline to enter into those they believe to be disadvantageous. Implicit in this freedom is the right of actors to set different terms for their interactions with different other actors, and to gather and act on whatever information they deem relevant to their decisions. For example, workers are free to sell their labor at different rates to different employers, based on their predictions of the pros and cons of different jobs, and information-technology firms are free to quote different prices for setting up and maintaining different clients' systems, based on their predictions of the different needs of those different clients. Fairness would require that insurers be free to quote different premiums for different customers, based on their predictions of the different needs of those different customers.

Admittedly, this line of reasoning establishes only a prima facie case in support of the claim that it would be unfair to allow insurers to take customers' genes into account in the setting of premiums. This is because, in my view, the free market is morally justified only insofar it produces benefits such as individual happiness (in the form of freedom of occupation, for example) and greater prosperity for society (by the workings Adam Smith likened to those of an invisible hand). Indeed, it seems very unpromising to set such consequentialist considerations aside and argue that any individual or firm has a natural right to freedom as extensive as that of actors in a free market; such rights, construed as natural rights, are more often assumed than given sound justifications.[15] And if such consequentialist considerations (rather than, e.g., natural-law ones) are the basis for any moral justification that can be given for the free market, then it is not unfair to place restrictions on actors' freedom when there is more to be gained from such restrictions than from the continued unfettered operation of the invisible hand. In principle, then, it would be possible to argue that it would not be unfair to prohibit insurers from taking customers' genes into account in the setting of premiums, if it could be argued that the consequences for society would be sufficiently desirable. In the next section, however, I argue that such a prohibition can actually be expected to have undesirable social consequences.

3. *Adverse Selection and Unintended Consequences*

Prohibiting genetic discrimination in health insurance would further impair the already troubled health care system in the United States by artificially compromising the one essential conditions for a well-functioning insurance market, namely, approximate symmetry of information between insurers and insureds. How an informational asymmetry would arise from such a prohibition is obvious: If genetic discrimination in health insurance is prohibited, then while individuals will be able to use genetic testing to gain tremendous amounts of information about their future health care needs, insurers will not have, or (which comes to the same thing) will be required to proceed as if they did not have, that same information. How such an informational asymmetry would threaten the operation of the health-insurance market is a complicated matter that can best be explored through a thought experiment involving a simpler kind of insurance than health insurance, such as term life insurance.[16]

Imagine a company called Yearly Life Insurance (YLI) that sells only one product: a life-insurance policy that takes effect on the day you purchase it, pays your beneficiary $1,000,000 if you die within a year, and expires otherwise. The price you pay depends solely on your age on the day you purchase the policy, and although you can renew your policy to ensure uninterrupted coverage, you have to pay a slightly higher premium upon each renewal, in accordance with your slightly increased risk of death with each passing year. YLI has thousands of customers, from teens to centenarians.

Now suppose that, due to new legislation, life-insurances companies are prohibited from discriminating on the basis of age. YLI responds by setting its single premium for every customer at the average of the premiums that its customers had been paying, so that its annual revenue remains the same after the law as before. Suppose that this new rate for everyone is $150, and that this happens to be what YLI had been charging sixty-two-year-olds when its premiums were based on customers' ages. Then YLI's product is suddenly a lot more attractively priced for people older than sixty-two, and a lot more unattractively priced for people younger than sixty-two. As a result, YLI gains older customers, and loses younger ones.

This, however, raises YLI's annual expenses, since older customers die more frequently than younger ones do. To stay solvent, YLI raises its premiums; the new price happens to be what YLI had been charging sixty-six-year-olds before the law took effect. At the new price, YLI's product will be too expensive for some customers, and the non-renewing customers will be disproportionately drawn from the younger ranks of YLI's customer base (since insurance appropriately priced for sixty-six-year-olds will remain a relatively good deal for customers in their seventies and eighties and so on). Again, to stay solvent while serving an older customer base, YLI raises its premiums, this time to the price it had been charging seventy-year-olds before the law took effect. Again, YLI loses younger customers and has to set its premiums higher still. After a few iterations of this cycle, YLI's extremely high premiums have cost it all but its oldest customers—the ones who had been paying such high premiums in the first place. All its other customers have no life insurance at all; they haven't fared any better with other companies, because other companies are going through the same transformation. The result is extremely expensive insurance for extremely old people, and no insurance for anyone else.

Obviously the issue of prohibiting genetic discrimination in health insurance is much more complicated than this simple example: Health insurance provides benefits in varying amounts throughout one's life, not just in a fixed amount upon death; it is sold in more and less comprehensive policies, many other factors than genes are considered in the pricing of health insurance, and in the United States it is mainly provided through employers rather than being sold directly to individuals.[17] Nevertheless, accounting for these complications leaves the story essentially unchanged when applied to health insurance in an age of genetic testing the results of which are kept hidden from insurers. Customers who anticipate relatively high health care expenses will find health insurance more of a bargain than those who anticipate relatively low health care expenses. As the former customers buy more insurance and the latter buy less, insurers will be forced to raise their premiums in order to stay solvent, leading to extremely expensive insurance for customers who anticipate extremely high health care expenses, and no insurance for anyone else.

This phenomenon, called adverse selection, is not merely a theoretical abstraction; its real-world importance was sufficient for the 2001 Nobel Prize in economics to be awarded to three economists "for their analyses of markets with asymmetric information."[18] Moreover, it can arise not only in insurance, but in any market in which buyers and sellers are asymmetrically situated with regard to information relevant to their transactions. (One of the three Nobel Prize winners is most famous for an article about asymmetric information in a used-car market.[19]) Nevertheless, its relevance to insurance is particularly acute. When customers have information that insurers do not have, or are not allowed to act on in the setting of premiums, the market tends to evolve so that nearly all customers get priced out of it. Prohibiting genetic discrimination in health insurance would establish an informational asymmetry between customers and insurers, an asymmetry that would currently be significant and that would be likely to grow only more significant as genetic testing becomes more sophisticated and widely available to individuals wishing to make more-informed predictions about their future heath care needs. It is ironic that the outcome that most advocates for a prohibition on genetic discrimination in health insurance are trying to prevent is one of decreased access to affordable health insurance, since that is precisely the outcome to which such a prohibition would tend to lead.

4. *Toward Systemic Reform*

I have argued that it is not unfair for insurers to distinguish among prospective customers on the basis of their genes, and it would not be socially desirable to prohibit insurers from doing so. On the contrary, such a prohibition would probably be counterproductive. Nevertheless, the goal of preventing decreased access to affordable health care is a worthy one, and if prohibiting genetic discrimination in health insurance will not achieve it, another means to it must be found. The obvious solution, in my view, is for the United States to institute a system of publicly funded basic health care for every resident, irrespective of genes and other indicators of future health care needs.[20]

In advocating this solution to the problem of potentially diminished access to affordable health insurance in an age of genetic testing, I am obviously offering nothing novel; I mention my view on this topic only to locate my position vis-à-vis two other positions regarding health care and health insurance. Some proponents of prohibiting genetic discrimination in health insurance reject publicly funded universal basic health care; they favor preserving the present system and they see prohibiting genetic testing as a means to doing so. Obviously I disagree with proponents of this position about both the desirability and the feasibility of preserving the present system.[21]

Other proponents of prohibiting genetic discrimination in health insurance agree that it is imperative for the United States to adopt publicly funded universal basic health care; they and I just disagree on what should be done in the absence of that first-choice solution.[22] But those who advocate prohibiting genetic discrimination in health insurance, but who agree that publicly funded universal basic health care is the real solution to this and many other problems, need to appreciate that by failing to acknowledge the probable consequences of prohibiting genetic discrimination in health insurance—namely, the problem of adverse selection discussed already—they are undermining the movement for the wholesale reform they agree is really needed. For they are implicitly suggesting that the present system, although imperfect, can be protected from the disruptions of genetic testing with the right kind of tweaking. In doing so, they overstate the promise of the present system and effectively lend support to those opponents of publicly funded universal basic health care who say that the stability and adaptability of the present system obviate the need for more thoroughgoing reform.

The ongoing revolution in genetics raises understandable fears about genetic discrimination. It is natural, when confronted with such a threat, to think of a prohibition as the way to stop it. Upon examination, however, such a prohibition does not possess either the moral warrant, or the economic rationale, that it might initially seem to have. Those who want to preserve broad access to affordable health care, and especially those who favor the adoption of publicly funded universal basic health care, should reject calls for a prohibition on genetic discrimination in health insurance.

Notes

1. *Gattaca*, written and directed by Andrew Niccol, released by Columbia Pictures (1997).

2. On this latter point, see Elliott Sober's appendix "The Meaning of Genetic Causation," in Allen Buchanan, Dan W. Brock, Norman Daniels, and Daniel Wikler, *From Chance to Choice: Genetics and Justice* (Cambridge: Cambridge University Press, 2000), pp. 347–370.

3. See Mary-Claire King, Joan H. Marks, and Jessica B. Mandell, "Breast and Ovarian Cancer Risks Due to Inherited Mutations in *BRCA1* and *BRCA2*," *Science* vol. 302 (October 24, 2003), pp. 643–646.

4. For nuanced discussion of this point, see Alice Wexler, *Mapping Fate: A Memoir of Family, Risk, and Genetic Research* (Berkeley: University of California Press, 1995).

5. See Buchanan et al. (cited in note 2), p. 27. See also Philip Kitcher, *Science, Truth, and Democracy* (Oxford: Oxford University Press, 2001), p. 184.

6. E. Virginia Lapham, Chahira Kozma, and Joan O. Weiss, "Genetic Discrimination: Perspectives of Consumers," *Science* vol. 274 (October 25, 1996), pp. 621–624.

7. Mark A. Hall and Stephen S. Rich, "Laws Restricting Health Insurers' Use of Genetic Information: Impact on Genetic Discrimination," *American Journal of Human Genetics* vol. 66 (2000), pp. 293–307.

8. Lisa N. Geller, "Current Developments in Genetic Discrimination," in Joseph S. Alper, Catherine Ard, Adrienne Asch, Jon Beckwith, Peter Conrad, and Lisa N. Geller, eds., *The Double-Edged Helix: Social Implications of Genetics in a Diverse Society* (Baltimore: Johns Hopkins University Press, 2002), pp. 267–285.

9. See, for example, Francis S. Collins and James D. Watson, "Genetic Discrimination: Time to Act," *Science* vol. 302 (October 31, 2003), p. 745.

10. Buchanan et al. (cited in note 2), p. 339.

11. In fact, Wisconsin's state-sponsored program for "Wisconsin residents who either are unable to find adequate health insurance coverage in the private market due to their medical conditions or who have lost their employer-sponsored group health insurance" is called the Wisconsin Health Insurance Risk Sharing Plan. See http://www.dhfs.state.wi.us/hirsp/index.htm, accessed on June 16, 2006.

12. Ironically (but fortunately from the point of view of social justice), the Wisconsin Health Insurance Risk Sharing Plan is only partially funded by actual risk sharing—i.e., policyholder premiums. In 2005, such premiums funded 57.9 percent of the program's costs of $174.5 million, with the remainder being funded by legally compulsory "assessments paid by insurance companies writing health insurance policies in Wisconsin . . . and reduced payments to providers." See "Health Insurance Risk Sharing Plan: 2005 Annual Report" (http://www.dhfs.state.wi.us/hirsp/reports/annual_2005.pdf, accessed on June 16, 2006), pp. 1–2.

13. This argument is suggested by the proposal that insurers should be legally allowed to set higher premiums for customers with preexisting conditions, but prohibited from setting higher premiums for customers with merely anticipated conditions for which they are currently asymptomatic. This proposal is attributed to Henry Greely, and discussed approvingly, by Deborah Hellman in her "What Makes Genetic Discrimination Exceptional?," in Verna V. Gehring, ed., *Genetic Prospects: Essays on Biotechnology, Ethics, and Public Policy* (Lanham, Md.: Rowman & Littlefield, 2003), pp. 85–97). It should be noted that this proposal, as described by Hellman, concerns legislation and not fairness per se.

14. Karen H. Rothenberg and Sharon F. Terry, for example, write that "It is only reasonable to be concerned that health insurers and employers may not fully understand the implications and limitations of genetic test results and the complex relationships between genotype and phenotype." See their "Before It's Too Late—Addressing Fear of Genetic Information," *Science* vol. 294 (July 12, 2002), pp. 196–197.

15. See, for example, Robert Nozick, *Anarchy, State, and Utopia* (New York: Basic Books, 1974).

16. The threat of adverse selection resulting from prohibiting genetic discrimination in life insurance is briefly alluded to by William Nowlan in "A Rational View of Insurance and Genetic Discrimination," *Science* vol. 297 (July 12, 2002), pp. 195–196).

17. For more on the role of employers in providing health insurance for most Americans, and for a brief history of the events that (partly accidentally) brought about the present system, see Henry T. Greely, "Health Insurance, Employment Discrimination, and the Genetics Revolution," in Daniel J. Kevles and Leroy Hood, eds., *The Code of Codes: Scientific and Social Issues in the Human Genome Project* (Cambridge, Mass.: Harvard University Press, 1992), pp. 264–280).

18. See http://nobelprize.org/economics/laureates/2001/index.html, accessed on June 16, 2006. Adverse selection is discussed in most microeconomics textbooks and economics reference works.

19. George A. Akerlof, "The Market for 'Lemons': Quality Uncertainty and the Market Mechanism," *Quarterly Journal of Economics* vol. 84, no. 3 (August 1970), pp. 488–500.

20. Although the Wisconsin Health Insurance Risk Sharing Plan (discussed earlier, in notes 11 and 12) is framed as an insurance program, the fact that the state funds more than 42 percent of its costs (by taxing health insurers selling policies in Wisconsin and requiring discounts from health care providers) makes this program a laudable step in the direction of the sort of solution I advocate.

21. An obvious defect of the present system, of course, is that it leaves more than one in seven Americans uninsured. See Carmen DeNavas-Walt, Bernadette D. Proctor, and Cheryl Hill Lee, *Income, Poverty, and Health Insurance Coverage in the United States: 2004* (an August 2005 report of the U.S. Census Bureau), p. 16.

22. See, for example, Philip Kitcher, *The Lives to Come: The Genetic Revolution and Human Possibilities* (New York: Simon & Schuster, 1996), pp. 132–139.

Drawing upon the Right to Privacy to Craft Laws to Protect Genetic Information

Julia Alpert Gladstone
BRYANT UNIVERSITY

The completion of the sequencing of the human genome and the resulting scientific advancements that have been made to enable ready access to information about a person's DNA present ethical challenges to society.[1] The public has a very complex relationship with genetics; we find that the gene is sacred, yet many of us fear its power. The allure of genetic science is reinforced by media, science, and the very legislation that is enacted to protect our personal interests in our deoxyribonucleic acid ("DNA"). There are numerous rationales for protecting information about one's DNA that have led to several legal approaches, which range from courts addressing DNA as personal property to antidiscrimination laws being introduced in state and federal legislatures. Our privacy is revealed in a unique fashion by the information in DNA, and therefore DNA ought to be respected as a part of human identity in the privacy protection paradigm.

The first part of this essay discusses the reasons why genetic information is viewed as special and how this argument has led to the acceptance of "genetic exceptionalism" by many policymakers.[2] The harms of treating DNA under this approach are briefly reviewed prior to an analysis of privacy that focuses on the dependent relationships of freedom and autonomy to privacy. The right to privacy as it has been interpreted in relevant United States courts' decisions is analyzed in Part II to describe a model preferable to genetic exceptionalism to protect information about one's DNA. Classroom topics and discussions are also explained in this section. Part III explains the essential nature of that right to privacy and suggests why a property model is not a useful model to follow. In the conclusion, I suggest that there are different approaches to protect DNA under the law that need to be streamlined along the privacy model of the tort of "misappropriation of identity."[3]

Part I. Is Genetics Information Special?

The study of the common patterns of DNA sequences in the human genome has enabled scientists to explore the relationship between information stored in genes and our phenotype. In addition, insights into the multiple genetic and environmental factors that cause disease have been made in a variety of genetics research programs that facilitate ways to improve human well-being. For example, association studies have lead to discoveries of the genetic risk factors for diabetes, cardiovascular disease, and cancer.

The media in our country, which have played a significant role in creating the DNA Mystique, are in part responsible for the public's misunderstanding of the importance of DNA. The notion that genes can determine and explain everything about us is very popular in our culture; the predictive nature of our genes is exciting. Headlines proclaim that much of our behavior is the result of our genes rather than the real and complex relation with the environment.[4] Science fiction in television and movies, in particular, has perpetuated the notion of the simplicity of tampering with DNA to achieve mighty results. The promise of cures for illness from genetic research has been deeply rooted in society as early as the 1980s.

The public's perception of the power of genetics is not all positive; the horrific memories of the eugenics abuses of Nazi Germany persist. They may in fact reinforce the present fears of genetic discrimination. Genetic information is like a "future diary," because it is highly predictive of the diseases one may develop. Many fear this information will be misused by employers and insurers. A related concern is that when discrimination is based on a gene that predominately affects a discrete group, racial, ethnic, or gender bias may be perpetuated.

The role of DNA as a unique identifier has been embraced for law enforcement and for similar detection purposes. Genetic information is highly personal, and it can be obtained without the person even knowing it has been taken. It has been described as very sensitive and stigmatizing, sometimes even referred to as a "scarlet letter."[5]

The scientific community has also contributed to creating the image of genetics as all-powerful. Understandably, scientists have faith in the importance of their work, and the fact that the U.S. government allocated three billion dollars to the Human Genome Project suggests its value. More interesting, however, is the 5 percent required allocation from the three billion dollar amount designated to explore the ethical, legal, and social implications (ELSI) associated with the genetic discoveries. This focus on ELSI scholarship suggests that genetics is special.

The threats of genetic discrimination that have been emphasized by the media perpetuate the need for a "genetics exceptionalism" perspective. Media-reported findings of genetic discrimination have been based on only a few studies with several methodological limitations, primarily that they are based on self-reported incidents of discrimination and anecdotal evidence. The public fears of genetic discrimination may prevent individuals from undergoing simple and valuable genetic testing or engaging in needed genetic research programs. The public health consequences of misplaced fears are real, as singling out genetics information for protection may create its own negative consequences.

There are powerful reasons for carving out special protections for genetics information, as explained earlier.[6] "Genetic exceptionalism" suggests that the laws will be established to treat genetics information as special in order to prevent discrimination based upon genetic information. Genetics exceptionalism has been shown to be a misguided theory because it leads to laws that are both overinclusive and underinclusive, as described later. In order to protect one's genetic information, DNA must be viewed as a key, but not exclusive, component to one's identity. One's freedom and autonomy are threatened by modern technologies that invade one's privacy. This invasion takes many forms, and DNA identification technology is among the most pervasive.

There is no specific genetics federal legislation, despite many attempts at passing a law.[7] The Health Insurance Portability and Accountability Act of 1996 ("HIPPA"), which offers protection by safeguarding all medical information under a privacy model, provides a better approach. Forty-six states have legislation that directly address the treatment of genetic information. They vary in their approach in that they may target general, health, or employment insurance discrimination based upon genetic information.

The underlying rationale against genetic exceptionalism is that it focuses on the gene as a characteristic of our physical makeup. Legislation that describes genes as special, unique, and separate from ourselves as whole beings sets up a paradigm where we do not control our genes. This is overinclusive because although it is true that we inherit our genes, receiving a diagnosis that we carry a gene for a particular disease is not determinative of our future health. For example, there are several preventive measures that one can take if one has the gene for colon cancer. In addition, someone may carry a single copy of a recessive gene, which may increase the chances of having an affected child but does not increase the future risk of diseases in the carrier.

The question of whether genetics is a proper proxy for what is not in our control leads to the harms of underinclusiveness as well.[8] There are many social and economic elements that are beyond our control and that may impact our lives as much as our genes, yet we do not justify protection legislation from socioeconomic status or culture. In addition, genetics is not alone in its predictiveness. Why should insurers be prevented from considering genes but not a significant history of exposure to harmful chemicals? These are some negative implications that result from treating one's genetic information as special and not simply as a component of one's identity under the general privacy umbrella.[9]

Part II. The Importance of Privacy

The discussion of privacy in the United States largely begins with the 1890 seminal *Harvard Law Review* article by Warren and Brandeis entitled "The Right to Privacy," where the authors set out the value of the "right to be let alone." It does not appear to be a static value; rather, as society has evolved, the concept of privacy in the United States changes, and the legal culture adapts accordingly.

It was in the case of *Katz v. United States*, 389 U.S. 347 (1967), that the Fourth Amendment was found to protect people, not places. The Fourth Amendment's protection against unreasonable search and seizure was found to protect a person speaking in a public telephone booth because he or she had an expectation of privacy. *Katz* was a case of the FBI's unauthorized use of a wiretap, which was a new technology at the time, and a petition to suppress the evidence that was obtained. Although Katz was in "plain view," what he sought to exclude "was not the intruding eye but the unwanted ear" (*Katz v. United States*, 389 U.S. 347). Working with the expanded definition of privacy, the *Katz* court states, "the average man [*sic*] would very likely not have his feelings soothed any more by having his property seized openly than by having it seized privately." *Katz* widened the scope of privacy from the secrecy paradigm developed from Warren and Brandeis's work by adding the two-part "expectation-driven" test.[10] First, the defendant must have an actual or subjective expectation of privacy; second, the expectation must be one that society is prepared to recognize as reasonable.

The use of the expectation of privacy test for defining privacy and its interplay with technology is well demonstrated in the court's analysis in *Kyllo v. United States*, 533 U.S. 27 (2001). Based upon the tip of an informant, law enforcement officers used a thermal imaging device to scan the defendant's home from across the street to reveal that portions of the house were unusually hot. Using the information from the imaging scan, the officers obtained a search warrant to find that the defendant was in fact growing marijuana in his house. The court held that the warrantless use of the thermal imaging device constituted an unreasonable search under the Fourth Amendment.

Justice Scalia writing for the majority held that the use of the thermal imaging device was a violation of the Fourth Amendment because the defendant had an expectation of privacy in his home, which this device violated. Scalia cited evidence that this technology is not in general use but suggested that when the device is commonly used a defendant will no longer have such an expectation of privacy in his home. An individual's freedom from unwanted interference in personal decision making has been reduced in recent years by technology. Encroachment comes from many sources. The current most controversial technologies involve data-mining systems, which consolidate both public and private information about individuals. People have come to expect an environment where privacy is nearly nonexistent; this surveillance has impacted individuals in many, sometimes hidden but harmful ways. Jeremy Bentham's design of prisons explains how society is put at risk by modern information-gathering techniques.

The Panopticon was a model prison designed by Bentham in the nineteenth century and was based upon the principle that one guard could watch many prisoners at the same time. Each prisoner would not know whether or when the guard was watching him because of the circular architecture of the building, the placement of windows, and the use of light. Consequently, each prisoner believed he was always being watched. The French philosopher Michel Foucault later drew upon the Panopticon as a metaphor for the modern mechanisms that are used to control society, showing its use in the design of factories, schools, and hospitals. Foucault and other philosophers who have used the Panopticon to describe the system of control that permeates society today recognize that the watchful eye will work even when there is no one in the guardhouse, which is similar to the internalization process of the reduced expectation of privacy that has over taken society today. Parties other than the government participate in the data surveillance and contribute to the awareness or sense that one is always being watched.

The expectation-driven conception of privacy is used by the courts both in finding an invasion of privacy tort and in defining a right to privacy in the Constitutional sense. Common-law precedent and statutes are guidelines, but the court will also look at contemporary community standards and become an arbiter of norms. Government, business, and other social institutions influence society's expectation of privacy, which has lead to an effective encroachment by incremental incursion into unsettled expectations, and once the incursions are in place they become internalized. Therefore, defining privacy according to expectations becomes a self-fulfilling prophecy outside of the control of the watched person.

The deciphering of the chemical structure of deoxyribonucleic acid (DNA) began with the work of James Watson and Frances Crick in 1953. Since the 1980s, the study of genomes has primarily been an international project aimed at obtaining a detailed map and gene sequence for a variety of organisms, culminating with the Human Genome Project (HGP) (http://www.genome.gov). The HGP is the term used to describe a collective group of projects from around the world organized in 1980 to create an ordered set of DNA sequences from known chromosomal locations and to develop new computational methods for analyzing genetic maps and DNA sequences. In June 2000 a working draft of the human genome was unveiled and on April 14, 2003, with 99 percent of the genome sequenced, the HGP announced that it had completed the sequencing of the human genome.

DNA is a double-helical molecule with four chemical components called nucleotide bases that are linked together into long sequences. There are only four distinct nucleotides that make up the double helix; these are adenine (a), cytosine (c), guanine (g) and thymine (t). It is the combination of these four bases—their sequence in a strand of DNA—that gives human life its character. A genome, therefore, is the continuous thread of DNA sequences on all forty-six chromosomes in the human cell. A gene is a length of DNA that contains sufficient information to code for, or to make a protein. The human genome contains approximately 20,000 to 25,000 functional genes (www.ornl.gov/sci/techresources/human_Genome/project/20to25K.shtml).

Bioinformatics provides the computer technology, methods, and algorithms for organizing these voluminous genomic data and their subsequent retrieval. In its simplest form, bioinformatics is based upon sequence homology between known and unknown sequences or proteins. Having the ability to search through a database of similar sequences and output a list of sequences that is statistically similar to the one already uncovered is a meaningful first step in the bioinformatics research process. The information generated by genomic research is growing exponentially and the cost to sequence an individual's human genome has dropped below $1,000.[11] As science advances, DNA can be used to predict susceptibility to particular diseases and reactions to medication, as well as predispositions to certain behaviors or sexual orientations. This would appear to increase the potential for genetic discrimination by government, insurers, employers, and others. It is the tremendous insights into the many intimate aspects of a person and their families that suggest the law must be developed to draw upon DNA as a key element of one's identity in order to assure adequate privacy protections.

Students at my university are always intrigued when I tie genomic topics into my regular legal studies classes, which I do on a regular and consistent basis as a successful method to teach ELSI topics to all my students. The issue of privacy is covered in a large number and variety of legal studies courses, and adding the dimension of genetic information to the analysis of what deserves protection under the "right to privacy" deepens the student's understanding of the U.S. definition of "expectation of privacy."[12] The general student population is usually familiar with some genome science, and I use this knowledge as the starting point of our discussion of the ethical implications of that discovery to one's privacy.

The Internet offers a rich resource of information for the level of instruction that I require to teach the ELSI issues, such that I change the required readings each semester; I usually assign one or two articles pertaining to a legal development in response to a genomic advancement. The discussion will inevitably broaden from the legal to the ethical and moral questions, where I offer as many alternative interpretations as I deem relevant.[13]

Part III. Privacy, Identity, and DNA

The passage by the federal government in 2000 of the DNA Analysis Backlog Elimination Act of 2000 ("DNA Act"), Pub. L. No. 106-546, 114 Stat 2726 (2000), which authorizes the extraction of DNA from those convicted of "qualifying federal offenses," furthers the privacy protection boundary that society or, more specifically, the legal system is creating for information about our DNA. The FBI is in charge of maintaining these

DNA profiles in the Combined DNA Index System ("CODIS"), which is a three-tiered system of national, state, and local databases that total more than 1.6 million DNA profiles. DNA profiles in CODIS are organized into two groups: the Offender Index, which contains DNA obtained under the applicable local, state, or federal law, and the Forensic Index, which contains DNA profiles obtained from the scene of the crime. Matches are made among and between the indices to help identify perpetrators of crime.

Short tandem repeat (STR) technology, which uses "junk DNA," is used to generate the DNA profiles for CODIS. At the present time only sex and race can be identified from this DNA. In 2000 a working group of the National Institute of Justice produced a report citing some concerns about the DNA collection, storage, and analysis. Along with other discrepancies, the group found that samples were being reanalyzed without consent. In addition, there was no clear overall policy as to what happens to DNA samples once they are added to the database, and many states do not require the destruction of a DNA record after a conviction has been overturned.[14]

All of the fifty states have statutes that allow some type of DNA collection from convicted felons to be included in the CODIS database; the role of DNA databases has been embraced as a powerful law enforcement tool, and while all states require DNA retrieval for sex offenders, there is a wide margin of variation between states as to their classification for others subject to inclusion. The U.S. Supreme Court has not addressed the constitutional legitimacy of requiring DNA collection from felons; state and federal courts around the country are upholding these statutes relying on one or both of the Fourth Amendment justifications described next.

The right to privacy is protected under the Fourth Amendment, and a reasonableness test based upon an expectation of privacy is used to determine whether a search has violated one's privacy. Looking at the "totality of the circumstances," the first justification balances the felon's severely reduced expectation of privacy against the government's overwhelming interest in identifying criminals and preventing recidivism. The second justification uses the "special needs" doctrine, which disposes of the reasonableness analysis to protect privacy rights when government intrusion serves special needs beyond normal needs. The "totality of the circumstances" and the "special needs" tests have given the green light to legislatures around the country to pass laws to collect DNA samples from individuals who have transgressed the law, and this has been sanctioned by the courts.

In 2003 the Ninth Circuit found that the DNA Act, which requires the DNA profiling of certain conditionally released federal offenders in the absence of individualized suspicion that they have committed additional crimes, violates an individual's Fourth Amendment right against unreasonable search and seizure, in *United States v. Kincade* 345 F.3d 1095 (9th Cir. 2003) ref'd *en banc*, 379 F.3d 813 (9th Cir. 2004) (6–5 decision). Judge Reinhardt, who authored the 2003 opinion, warned of the dangers of storing information about our citizens in a centralized place and the risks associated with having DNA samples on "file in federal cyberspace." This decision was reversed in April 2004, however, by an *en banc* decision of the court, which aligned itself with the Fourth Amendment justification arguments of prior judicial decisions. The Ninth Circuit's reversal of itself in the *en banc* decision of *United States v. Kincade*, 379 F.3d 813 (9th Cir. 2004), reflects the controversy and uncertainty surrounding the legal definitions of one's self and the permissible extent of the government thereon.

Judge Kozinski, writing for the dissenting minority in the final decision, states that when the time comes to create a DNA database for the general public the judges say that they will "stand vigilant and guard the line, but by then the line, never very clear to begin with, will have shifted. The fishbowl will look like home" (*United States v. Kincade*, 379 F.3d 813, 873 [9th Cir. 2004]).[15] This statement emphasizes the view that our expectations define the legal boundaries of privacy and therefore we must be aware of the central role that privacy plays and treat it with respect under the law. Judge Kozinski's insight also shows that the judicial system does place a high value on information about our DNA because it is only under "special needs" or using the "totality of the circumstances" that the government can have access to DNA profiles.

Judge Reinhart, also writing for the dissenting minority, focuses on CODIS's potential to expand, which can cause severe injuries to basic human freedoms. The DNA "fingerprint" that is now entered into CODIS is a "sleeping tiger" for all the information that it may eventually be able to reveal. The "junk DNA" of today may some day be used to show the presence of traits for thousand of diseases or behavioral characteristics. The minority dissent of five made it clear that "the permanent maintenance of this (DNA) information about untold millions of Americans, intrudes into the core of those intimate concerns which lie at the heart of the right to privacy" (*United States v. Kincade*, 379 F.3d 813, 848). The value of protecting DNA is reflected nicely by the Benjamin Franklin quote at the introduction of the dissent: "They that can give up essential liberty to obtain a little safety deserve neither liberty nor safety" (*United States v. Kincade*, 379 F.3d 813, 843).

At the present time the distinctions that determine one's privacy expectations under the law and the legal treatment of the DNA information suggest that a general population DNA database would be regarded as an affront to human dignity.[16] If one is fearful of the "slippery slope" of legal and social trends expressed by Judge Kozinski, however, it is necessary to regard DNA as an integral element of one's identity and to connect one's need for autonomy to the right to privacy under the law.

The personal identity of an individual implicates his or her concerns for privacy, and as biotechnology and similar emerging technologies make the knowledge of DNA readily and commercially available, certain legal authority has focused on property rights in the body and informed consent as a means to protect one's interests. Intellectual property law has been applied to grant patent rights on DNA sequences, which suggests that there are property interests in DNA information (an explanation of this area of the law goes beyond the subject of this article). In 1990 the California Supreme Court was confronted with the question of whether John Moore had a property interest in the valuable pharmaceutical products that had been derived from his spleen cells. In a remarkable decision that has served as precedent in the field ever since, the court overturned a lower court ruling to find that Moore did not have a property-based right to share in the profits derived from the use of his cells. His cause of action for conversion was denied (*Moore v. Regents of the University of California*, 793 P.2d 479 [Cal. 1990]).

The California Supreme Court denied Moore's claim for appropriation based upon his DNA because it found nothing unique in the particular cells that were derived from Moore's spleen: "lymphokines, unlike a name or a face, have the same molecular structure . . . in every human being and the same, important functions in every human being's immune system" (*Moore v. Regents of the University of California*, 793 P.2d 479, 490). Further, the court accepted the argument that it was only through the efforts of the defendants that the spleen cells were made into patentable blood cell lines.

Both the Appellate Court and the Supreme Court were concerned with questions of dignity that arise in the commercialization of human body parts. Reminiscent of the progress of humans in society, allusions to slavery come to mind whereby one person becomes the property of another. The appropriation cause of action that Moore relied upon focused on the body as physical property with a financial value. The court rejected this, finding that the physical human body as distinguished from the mental and the spiritual has little or no marketable value. By focusing on the proprietary interest in the DNA as raised by Moore himself, the Supreme Court did not find a legitimate interest. The California Supreme Court found in Moore's favor but based upon breach of fiduciary duty and informed consent.

The California Supreme Court's decision in *Moore* suggests that genetic information is not a property interest and should not be treated with special or more protections than any other confidential information. Laws that protect information only because it is genetic can detrimentally alter the social meaning of a person. Laws that encourage genetic essentialism suggest that a person's DNA is the essence of the person, rather than regarding genetic information as one attribute among many including sense of humor, intelligence or societal influences. (See also Part I.) If we apply the principles of genetic exceptionalism we sacrifice our free will, altering our notion of identity. We can protect genetic information under a privacy model by including DNA in the general definition of one's identity without adopting a genetic exceptionalism belief system.

As discussed previously in the Panopticon analysis, surveillance technology is making a profound and harmful impact on an individual life because it robs one of privacy and freedom. DNA as a valuable part of one's identity and requires protection under the laws of privacy. Privacy protections must be used to protect DNA because without privacy one will lose one's sense of identity. There is a compelling argument that underlies the application of privacy protection to protect information about one's DNA, an argument that is founded on the principle that privacy is understood to be an aspect of human dignity.[17] Throughout society and in Europe particularly, privacy rights are recognized and respect for human dignity is taken seriously; the very essence of the European Convention on Human Rights is respect for human dignity and freedom.

Human dignity is an elusive concept, but it can be well explained in terms of autonomy, which is the personal power or ability that makes judgments and actions one's own. Autonomy of will bestows respect on an individual as an agent to make free choices, and as such it allows one the context and conditions in which to operate in order to engage in the conscious construction of the self. Autonomy embodies the understanding that one is a source of free and informed choice; this is an empowered view of human dignity.

The philosopher Kant ties this theory of autonomy to what is often referred to as the "Golden Rule," which is the principle that every human being has a legitimate claim to respect from his or her fellow human beings and is in turn bound to respect every other. Kant's philosophy extends beyond the nature of autonomy to its value, when he articulates that "Humanity itself is a dignity; for a human being cannot be used merely as a means by any human being . . . but must always be used at the same time as an end." Kant contends that one's autonomy is the sole principle of morals and therefore one's dignity comes from being morally autonomous. It is quite common to apply Kant's theory in support of the position that autonomy or free will is what is intrinsically and universally distinctive about humans. The essence of free will or autonomy to humanness, however, can be developed independently of Kant's deontological constraints.[18]

A less restrictive view of the development of one's autonomy depends upon an awareness and understanding of oneself as a sentient being. The void of privacy in our society as depicted in the Panopticon metaphor robs individuals of this capacity. The development of a self is eloquently described by John Stuart Mill in his *Autobiography* as a departure from the rationalist habit of analysis and fatal questioning, to the cultivation of feelings. Mill, who spent his childhood and early adult years engaged as a brilliant student of philosophy, pedagogy, math, and military and penal practices, describes his early adult life as being "stranded at the commencement of my voyage, with a well equipped ship and rudder, but no soul; with any real desire for the ends which I had been so carefully fitted out to work for." Mill had no affect and described himself as emotionally dead because he had never taken the time to develop an independent autonomous self. He had followed his father's views on life, which led him to an existential crisis. Mill writes that he acquired a habit of leaving "my responsibility as a moral agent to rest on my father, my conscience never speaking to me except by his voice." Mill discovers his self through poetry, which awakens feelings and the power of imagination. Through the cultivation of this inner life he is able to act autonomously and contribute to his community and society.

The tort of "appropriation of identity" was discussed by both the Appellate Court and the Supreme Court in the Moore case; the Supreme Court overturned the case because it could not see that the Appellate Court recognized and placed value on the dignitary aspect of the identity—rather, the Supreme Court references were mainly to the pecuniary returns to Moore.[19] A review of American jurisprudence suggests that American law does recognize a person's "identity" in terms of offending one's dignity; it is something that can be stolen or appropriated. In the case of *Pavesich v.*

New England Life Ins. Co., 50 S.E. 68 (Ga. 1905), the court found that the unauthorized publication of Pavesich's picture for commercial purposes deprived him of control over his identity, in effect "enslaving" him by forcing his image into the service of another against his will.[20] The knowledge that one's features are being used for such a purpose brings a person to realize that his [*sic*] liberty is being taken away from him; he is under the control of another and he is no longer free (*Pavesich v. New England Life Ins. Co.*, 50 S.E. 68 [Ga. 1905]). Although the law has long recognized "the tort of misappropriation of identity," it has been overshadowed because the tort of "the right of publicity," which is very similar, reaps greater financial rewards.

Conclusion

Drawing upon the tort of "appropriation of identity," one can assert that DNA is a constituent of individual identity which deserves legal recognition and protection. The Appellate Court in *Moore* in fact found that "a patient must have the ultimate power to control what becomes of his or her tissues. To hold otherwise would open the door to a massive invasion of human privacy and dignity in the name of medical progress" (*Moore v. Regents of the University of California*, 249 Cal. Rptr. 494, 508 [Cal 1990]). The identity interest at stake is not one of a pecuniary nature, which misled the California Supreme Court in Moore, but it is concern for the "human whole, the body, mind and spirit are one." The Appellate Court considered Moore's dignitary interests integral to his appropriation claim when they characterized his harm as a "massive invasion of human privacy and dignity" (*Moore v. Regents of the University of California*, 249 Cal. Rptr. 494,508 [Ct App.1988], review granted and superseded by 763 P.2d 479 [Cal 1990]).

The ease with which we can access information about our DNA has changed our view of ourselves and calls for a legal response because we want to protect our identity.[21] The expression of DNA, which historically had been unknown and which plays a part in determining our identity, impacts our human values and rights. Several legal approaches based upon various privacy models have been used to protect an individual's interest in his DNA information. We need to view the human body as the physical and temporal expression of the unique human persona; this will lead to a comprehensive view of human identity. Appropriate attention will then be placed on human identity as including DNA along with the spirit and the mind, as needing protection from the invasive technologies that threaten our right to privacy.[22] With such an approach, the tort of "appropriation of identity" can be used to protect DNA under the right to privacy without engaging in "genetic exceptionalism."

Notes

1. The International HapMap Project, 426 *Nature* 789 (2003).

2. D. Nelkin and S. Lindee, *The DNA Mystique, The Gene as a Cultural Icon* (New York: W. H. Freeman, 1995).

3. Harris Poll, 1986: 66 percent of respondents believed that genetic engineering would improve their lives; see Suter, S., The Allure and Peril of Genetics Exceptionalism: Do We Need Special Genetics Legislation? 79 *Wash. U. L. Q.* 669 (2001) footnote 12.

4. G. Smith and T. Burns, Genetic Determinism or Genetic Discrimination? 11 *J. Contemp. Health L. & Pol'y* 23 (1994).

5. H. Greely, Genotype Discrimination: The Complex Case for Some Legislative Protection, 149 *U. Penn. L. Rev.* 1483 (2001).

6. See infra Part II and III.

7. Genetic Information Nondiscrimination Act of 2003, S. 1053, 108th Cong. 202 (2003).

8. The American with Disabilities Act (ADA) is another federal law based on the discrimination model that has been used to protect individuals based upon their genes.

9. S. Suter, The Allure and Peril of Genetics Exceptionalism: Do We Need Special Genetics Legislation? 79 *Wash. U. L. Q.* 669 (2001).

10. L. Brandeis and E. Warren, The Right to Privacy, *Harvard L. Rev.* 1890.

11. B. A. Binzak, How Pharmacogenomics Will Impact the Federal Regulation of Clinical Trials and the New Drug Approval Process, *The Food and Drug Law Journal* 58, 103 (2003).

12. P. J. Gardner, U.S. Intellectual Property Law and the Biotech Challenge: Searching for an Elusive Balance, *The Vermont Bar Journal and Law Digest* 29 (2003).

13. Richard Cole, Authentic Democracy: Endowing Citizens with a Human Right in Their Genetic Information, 33 *Hofstra L. Rev.* 1241 (2005).

14. Mark Rotenberg and Marcia Hormann, EPIC AMICUS Brief in the Rehearing *en Banc* in the Ninth Circuit of the United States v Kincade, 379 F. 3d 813 (2004).

15. Idem.

16. Meghan Riley, Comment: American Courts Are Drowning in the "Gene Pool": Excavating the Slippery Slope Mechanisms Behind Judicial Endorsement of DNA Databases, 39 *J. Marshall L. Rev.* 115 (2005).

17. James Nehf, Recognizing the Societal Value in Information Privacy, 78 *Wash. L. Rev.* 1 (2003).

18. Roger Brownsword, An Interest in Human Dignity as the Basis for Genomic Torts, 42 *Washburn L. J.* 413 (2003).

19. Julie Cohen, Examined Lives: Information Privacy and the Subject as Object, 52 *Stan. L. Rev.* 1373 (2000).

20. Immanuel Kant, *The Metaphysics of Morals 209* (Mary J. Gregor, trans. & ed., Cambridge University Press, 1996) (1797).

21. Jonathan Kahn, Biotechnology and the Legal Constitution of the Self: Managing Identity in Science, the Market and Society, 51 *Hastings L. J.* 909 (2000).

22. Idem.

The Human Genome Project

Implications for the Study of Human Evolution

Alexander Werth
HAMPDEN-SYDNEY COLLEGE

When considering the Human Genome Project (HGP) and its many implications—scientific and technical, as well as ethical, legal, and social—most attention has focused on medical applications and related health care/biotechnology concerns. Due consideration has been given to identifying, mapping, and treating genetic ailments, and to such potential problems as discrimination in provision of treatment or insurance coverage. Less attention has been paid to the "human" part of the HGP in terms of how it defines and describes us as a species, both with regard to relationships within *Homo sapiens* and relative to other species. Not only does genomics hold great scientific potential for the study of human evolution and indeed all biological evolution—witness J. Craig Venter's recent initiative to discover and describe global biodiversity[1]—but as I show in this chapter, there are profound ethical, legal, and social implications for the study of humanity and its diversity.

The HGP is regarded as a resounding success for science, technology, and commerce despite the continuing absence of certain practical applications initially hyped from the project's launch in 1986 to publication of the genome's "first draft" in 2003. Although many such applications have not yet been realized, the numerous potential benefits of this immense bioinformatics project continue to accrue. Typically these fall into two categories: biomedical and purely scientific. Surely some scientific findings without tangible benefits today may lead to practical applications tomorrow, so that this boundary is more a blurred line than a clear distinction. Nonetheless, we can outline current and prospective biomedical applications, such as DNA-based testing to detect gene mutations and accurately diagnose resulting conditions, including predisposition to various forms of cancer; development of technology to isolate and clone genes responsible for human genetic disorders, resulting in better medical management; and pharmacogenomic development of "tailored" drugs. We can foresee additional outcomes of the HGP with no immediately apparent impact on health care or the criminal justice system, to name another practical beneficiary. Major gains in this "pure science" category involve insight into basic biology, including study of the organization of genes within genomes, comparison of genomes of different species, and identification of genes essential to specific developmental processes. These and other investigations will unquestionably aid in the study of molecular evolution, including evolution of our own species and our hominid ancestors.

However, my aim in this chapter is only partly to enumerate and elaborate on the HGP's implications for the study of human evolution. In addition, I wish to demonstrate that the widely recognized ethical, legal, and social implications of the HGP apply not only to biomedical situations, as is widely recognized, but often to "purely scientific" findings as well. The so-called "ELSI" considerations typically relate to health care, reproductive technology, and associated economic concerns, but just as the HGP provides a better understanding of hominid ancestry and relations, it yields information that bears on societal issues related to ethnographic diversity and, for example, the traditional practice of classifying humans by racial or other anthropological groupings. Hence this chapter concerns implications for but also of the study of human evolution.

In 1909 Wilhelm Ludvig Johannsen coined the term *gene*, now one of the most common and important words in biology and in society at large. The word derives from the Greek *genos*, referring to origin, as well as to generation, race, and clan; hence it is clear that "gene" refers not merely to particulate inheritance and snippets of DNA, but to us: human beings. We can study genes of mice and yeast, but we are naturally most interested in human applications and human history. Our fascination and preoccupation with our own origins predate Darwin, and now, with the advent of modern genetics, we can study the past, present, and even the future of humanity not merely with dusty bones and stone axes but with the precision of high-tech lab technology. Human evolution is of course a volatile and timely topic. One can hardly pick up a newspaper or turn on a television these days without spying a reference to the "intelligent design" creationism controversy in school districts and state legislatures. The knowledge that we gain from the HGP is therefore of interest not only to professional anthropologists but indeed to all of us, as it bears on ubiquitous societal issues of ethnography and race. There are obvious repercussions for how we deal with this information, and for how we use it to treat our fellow humans and even, some would say, our nonhuman relatives.

Genetics continues to change the face of anthropology by providing dual perspectives on phylogenetic relationships: inter- as well as intraspecific, elucidating our relation to nonhuman (including ancestral) species as well as revealing the complementary spectrum of diversity within *Homo sapiens*. Genomics allows us to study haplotypes and single-nucleotide polymorphism (SNP), providing insight into human biogeographic distribution and historical migration patterns and the origin and spread of inherited disorders. Understanding our evolutionary history better allows us to contemplate genetic screening and testing as well as eugenics and genetic "enhancement." Numerous other ELSI concerns relate to human evolution, including access to genetic information by researchers. Patenting of genes and GM organisms likewise impinges on evolutionary study, with potential ethical consequences for how we ought to treat nonhuman species. Finally, just as the HGP informs our study of evolutionary history, it simultaneously provides a glimpse of our likely evolutionary future. For example, use of reproductive or therapeutic cloning, preimplantation genetic diagnosis, stem cells, and genotypic cures (as opposed to phenotypic treatments) involving gene splicing might all be seen as parts of a further stage of human evolution that allow genetic recombination or other alteration without mating.

The story of human evolution is not merely about australopithecines, stone tools, and the switch to upright, bipedal locomotion. Evolution is not simply a historical record we can evaluate in retrospect, but an ongoing process of changing gene frequencies within populations. Those changes leave a record in the genetic code that can be investigated in the way that morphological and behavioral changes are studied through fossils and implements. Our genetic material resolves issues relating to large-scale or "macro-evolutionary" change (relation to other species, living and extinct) as well as "micro-evolutionary" change within *Homo sapiens*, as distinct human populations accumulated genetic markers linked to adaptive changes relating to particular environments as well as neutral changes that simply tell a story of which groups interbred with others, and roughly where and when genetic reshuffling and divergence occurred.

What has the HGP told us so far, and what is it likely to reveal in the future? We know, based on genetic as well as physical and archaeological evidence, that we are all descended from a relatively small group of people who lived in Africa no more than 150,000 years ago (about 5,000 generations ago), according to current estimates, and more likely only 60,000 years or 2,000 generations ago.[2] We know that in succeeding generations humans dispersed and evolved largely in isolation from each other in extended family clans or tribes, and that many metabolic and external features evolved rapidly in each population for adaptive reasons or due to sexual selection. Populations remained largely distinct until large-scale gene flow began with great voyages of discovery over land and sea, and of transport of large numbers of enslaved peoples, in the past millennium. We know that we continue to evolve with ever more rapid gene flow. A prominent geneticist once remarked that the bicycle was one of the most significant inventions of the twentieth century in that it greatly fostered gene flow, as people could quickly and easily marry and spread genes through a much wider area.[3]

The overall picture shows a species that exhibits extraordinary variability. A recent study[4] illustrates ways in which humans are still evolving, with genetic variants that may have been favored during major shifts over the past 10,000 years, such as the advent of agriculture, climatic change, and changes in habitat and diet. University of Chicago researchers scanned the human genome (using 209 unrelated individuals) for regions displaying high degrees of homogeneity, indicating low recombination and, presumably, signals of "recently adaptive" genes, such as the salt-sensitive hypertension gene. The strongest signal noted was for a mutation in the lactase gene, which allows adult humans to digest milk; this is prevalent in Europeans but very rare in people from other regions. Many of the genes noted in the study relate to carbohydrate metabolism, alcohol digestion, and metabolic rate (such as the conceptual "thrifty gene," which suggests changes in how the body stores calories), all perhaps stemming from dietary changes. Genes relating to skin pigmentation, reproduction, and sexual competition also seem to evolve quickly in humans, as in other primates.

The face of anthropology is shifting rapidly as the field increasingly relies on genetics. Information from the HGP is now used to answer such crucial questions as where, when, and how ancestral humans arose, and with what relationships and diversity. Just a generation ago we learned of Mitochondrial Eve, the most recent common matrilineal ancestor of all living humans. Molecular clock techniques that correlate elapsed time with observed genetic drift have likewise proven useful in Y chromosome analysis, supporting the claim that modern humans originated in Africa and tracking the spread of peoples throughout human prehistory. Y chromosome analysis[5] has been used to trace the dispersal of Anglo-Saxons, Celts, and Mongols, and to investigate the fecundity of such well-known historical personages as Genghis Khan and Thomas Jefferson. Hence genetic genealogy is of use not only in modern paternity cases, but can, for example, help individuals and groups to determine ancestral origins and homelands, find geographic regions and living relatives for further genealogical study, and confirm or deny ancestral connections between small families or large populations. The National Geographic Society recently announced a five-year plan, dubbed the "Genographic Project," to collect and analyze DNA samples from more than 100,000 people worldwide to trace human origins and migrations. According to program director Spencer Wells, DNA is a "time machine" that will help researchers compile the most comprehensive database of anthropological information to answer basic questions of who we are and how we got here. We will be able to pinpoint where and when modern humans originated in Africa and how they diversified as they spread across the globe. As Wells notes, in spite of our tremendous diversity we are all "effectively members of an extended family."[6]

We now study the evolution of genes and gene families (such as globin genes) with potentially immense medical benefits, but we also address basic ideas about evolutionary theory. For example, does the human genome show evidence of slow, steady evolution or punctuated equilibrium?[7] What role does the preponderance of seemingly unused "junk" DNA play? What about "stealth" DNA? How do repeating DNA sequences (entire genes or coded exons) arise? How commonly do genes "jump" around the human genome, by either standard recombination or other methods of translocation? Continued study of the human genome could teach us as much about the general rules of molecular evolution as about specifics of human physiology and inheritance. A key term in modern genetic anthropology is *coalescence time*, the number of generations between the present time and a presumed common ancestor. This concept bears heavily in contentious debates such as the precise timing and geographic location of the first modern humans as well as of smaller populations such as Armenians or Ainu. Did *Homo sapiens sapiens* arise and disperse according to the Out-of-Africa model, as is now widely believed, or in separate populations around the Old World? When did humans reach the New World, and by what route? How closely (or distantly) related are various groups within Africa?

Genomic information reveals new data, as well as confirming and strengthening current theory. In just the past year, studies have indicated that native North Americans are descended from as few as two hundred original settlers; that genes active in the development of human brains and language abilities (e.g., microcephalin, FOXP2) arose as recently as 40,000 years ago and continue to evolve[8]; and not only that autism, schizophrenia, baldness, and thrill-seeking are linked to specific gene loci, but that we can now investigate how and when such traits arose and spread within various populations. Genomic studies also reveal the role of the environment—that is, what genes *don't* do—in human development and evolution.

Scrutiny of fossils and other relics will continue to play a vital role in paleoanthropology[9]; genetics and genomics will never wholly replace dental, cranial, and skeletal reconstructions. Consider how much we have learned in the past few years from Washington State's Kennewick Man and Indonesia's tiny, hobbit-like *Homo floresiensis* (both still in dispute). However, in the same few years we have gained a better understanding of how we relate to Neanderthals not merely from osteological analysis or exploration of recovered artifacts, but by testing DNA. Controversy remains but has abated greatly as evidence strongly suggests there was no interbreeding between *Homo neanderthalensis* and modern humans, and that the two merit placement as separate hominid species. In 2006, exactly 150 years after the discovery of the first Neanderthal fossils in Germany's Neander Valley, a team led by Svante Pääbo at the Max Planck Institute for Evolutionary Anthropology announced the launch of the Neanderthal Genome Project, aimed at recreating the entire DNA sequence of our closest relative, which went extinct roughly 30,000 years ago. The task of extracting DNA from ancient fossils (initially using 45,000-year-old bones from Croatia) is daunting, but Pääbo and other scientists hope that by comparing human and Neanderthal DNA they can discern key turning points in our evolution, distinguishing, perhaps, what enabled us to survive while our distant cousins were less successful.[10] Results of this two-year initiative could address the recurrent question of whether Neanderthals interbred with *Homo sapiens* (and if so, to what extent), as well as shed light on other aspects of their biology. Did they possess genes known to play a key role in language? What were their brains like? What was their skin color? How large was their total population size?[11]

Use of DNA evidence from this and other fossil material, especially when coupled with genomic study of modern humans and other apes, is likely to impact further our understanding of species lines and relationships among human ancestors. Haplotype mapping not only aids in answering questions of human history, it informs us about prehuman ancestry, including fossil hominids as well as living apes. We have all heard that we share over 98 percent of our DNA with chimpanzees (*Pan troglodytes*). Where and when did these lineages separate? Where do pygmy chimps (bonobos, *Pan paniscus*) fit in? What does the small amount of differing DNA relate to? Should humans and chimpanzees be placed together in a single genus (*Homo* or *Pan*), as some primatologists argue?[12]

A remarkable example of DNA's potential utility in elucidating human evolutionary history was provided by a recent study by the Harvard- and MIT-based Broad Institute, directed by famed geneticist Eric Lander, a leader of the HGP. Their much-heralded yet contentious genomic work[13] suggests our species's origin was much more complicated, lengthy, and recent than previously thought. The study mined a trove of newly available genetic data to compare human and ape DNA, relying on the basic molecular clock concept by which phylogenetic events can be dated based on the inexorable and predictably steady accumulation of mutations. Its results challenged the conventional timeline of human evolution, suggesting that ancestral proto-humans split off from apes about five million years ago (mya). Examination of the Toumaï specimen of *Sahelanthropus tchadensis*, thought to be the oldest fossil in our human family tree, had suggested a divergence date of roughly 6–8 mya, consistent with earlier molecular clock estimates. The reliability of molecular clocks is often called into question, and it is desirable that genetic data be integrated with morphological data. It is also possible that key hominid remains have been misdated, leading to confusion. However, this study also made headlines by suggesting not just when but also how the split occurred, yielding the provocative conclusion that the distant ancestors of humans and chimpanzees interbred long after they had split into two distinct lineages, and hence that the origin of humans was perhaps not a crisp breakup but instead a long, messy affair.

What made the Broad team's analysis more detailed than previous investigations, and their results presumably more precise, is that instead of merely looking at an average genetic difference between human and chimpanzee DNA, they calculated the difference at many different yet corresponding loci throughout the genome of a number of ape and other primate species, aligning matching DNA segments (roughly twenty million base pairs[14]) by computer and examining 800 times more genetic material than previous studies.[15] After comparing small sections of the genetic code, the team calculated, via sophisticated statistical techniques, a divergence time—how long it would have taken to accumulate all differences—for each pair of segments. The surprising results included strange genetic patterns, such as widely varying divergence times (by millions of years) for different sections of DNA, that could best be explained by "genetic collaboration" (i.e., interbreeding) between human and ape ancestors. Divergence time hence depends upon where in the genome one looks. Also surprising were data from the X chromosome, which consistently yielded more recent divergence times. This too could best be explained by interbreeding given that the X chromosome is a focus of hybrid fertility problems, such that X chromosomes in a hybrid population are thought to match one of the parent species. Interbreeding would place strong selection pressure to eliminate traits that contribute to X-linked hybrid sterility and inviability, and this could explain why human and chimp X chromosomes are so similar.[16]

The high degree of variation between genes led scientists to the controversial conclusion that the parent population first split into two lineages, one leading to humans and the other to chimps, around ten million years ago, but that after evolving apart for as much as four million years they reunited for a "brief fling" that produced a third, hybrid population with characteristics of both lineages.[17] Some genes differ substantially between the two modern species, such that they must not have mixed over the past ten million years, whereas other genes are much more similar, indicating contact no more than six million years ago. During the long period during which the populations may have interbred, they could have produced sterile offspring as well as those that remained fertile and hence created a mixed group, a dead-end lineage combining both prehuman and prechimp genes and features, to which the Toumaï individual belonged. According to the Broad team, one of these interbred hybrid groups ultimately began the human lineage. The young age of the X chromosome has been tagged a "smoking gun" for hybridization, but much more work is needed to support this intriguing hypothesis. Completion of gorilla genome sequencing by late 2007 should allow for a richer analysis, further revealing whether the evolution of certain genomic regions at different rates is indeed due to hybridization or to other mechanisms such as conserved functionally important regions or the presence of mutation hotspots.[18]

What is certain is that the origin of our species from ape-like ancestors, a thought that has always repelled many people, might have been, if it indeed involved interbreeding, an even more scandalous scenario, and was clearly a more complex event than even anthropologists had presumed. Note that this is not suggesting that modern humans and chimps interbred, but rather that our distant, long-extinct ancestors might have. Despite what much of the public mistakenly assumes, humans did not arise directly from chimpanzees, nor are chimpanzees "trying" to become human; rather, they are our closest *living* relatives, for we are both descended from a relatively recent common ancestor. Many extinct apes, including numerous hominid species, lived in the millions of years since our divergence. The Toumaï fossil, which combines primitive ape-like and derived human-like features, has been dated at seven million years ago, prior to what the Broad team's results suggest as the final split. If dated correctly, this fossil's human-like features suggest the human–chimp divergence may have occurred slowly, over a long period of time, and that some of the cranial characteristics we take to denote proto-humanity, such as large brains, may have preceded the clear, independent existence of our human lineage. Paleoanthropologists who were not part of the study agree that genetic analyses will force a reappraisal of human origins, especially given the incomplete fossil record.[19] At the very least we now know that the speciation event that ultimately led to the appearance of *Homo sapiens* was an unusual one, a story of monkey business indeed. In the words of University of California (UC) Santa Cruz geneticist David Haussler, "It's a fascinating tale of ancient shenanigans. I continue to be amazed at what DNA analysis can reveal."[20]

Not only will this study refine conventional thinking about human origins, but it will spur reevaluation of origins of all species. It is one of a wave of recent projects showing the difficulty of drawing precise lines between closely related species, particularly during divergence. It will prompt broader evolutionary study of interspecific hybrids in speciation, previously thought to be virtually nonexistent in animal species although widespread in plants. The Broad team, among others, will continue to examine genomic data to determine whether species, including our own, split apart abruptly or via a longer, fuzzy period of overlapping hybridization, potentially bringing together the best attributes of both parent species. Interbreeding is known to produce strong selective pressure for evolutionary change, and exchange of genetic information may be common as new populations emerge and diverge. In a striking parallel, another recent study suggests modern human genes reveal similar interbreeding among *Homo sapiens*, *Homo erectus*, and other hominids from 1.5 million to 80,000 years ago. In the words of population geneticist Alan Templeton, "We don't have a tree of human populations with branches for Europeans and Asians and Arabs. It's more like a trellis: things are intertwined."[21]

The Broad team's report could also help settle a controversy suggesting a higher mutation rate among males in primates than in other species.[22] Prior comparisons of sequence divergence rates in human and rat autosomes and sex chromosomes suggested males had a mutation rate twice that of females, but the human–chimpanzee analysis revealed a male mutation rate six to seven times higher. The low divergence found between human and chimp X chromosomes in the recent study suggests the formerly anomalous human–chimp data match human–rat findings for male mutation rate. Again, comparative study of the human genome yields clues not only about our own species but to the mechanisms at work in the origin and evolution of all life. We may study ourselves for selfish reasons, to gain practical biomedical applications and pure scientific knowledge about humans, but along the way we inevitably end up learning about the fundamental processes that govern all life, just as we learn that we are inextricably tied to all other life forms. This is of course never a bad thing to be reminded of.

Again, such reflection is not merely academic. This information carries profound ethical, legal, and social implications. I refer not only to the legal status of apes, which, if we expand our moral circles (à la Peter Singer) beyond humanity, are the next species in line for protection. How we treat our closest relatives is an important concern, but how we see ourselves, and our place in the scheme of nature, is an equally weighty issue that strikes at the heart of fundamental religious and philosophical questions that concern us all. The search for cures may have driven the HGP, but when we step back and view the bigger picture we see that the knowledge gained by this vast scientific enterprise has other applications, less tangible but no less momentous. Who we are and how we got to be this way are basic questions people have contemplated for millennia. Thanks to the HGP we have much more fodder for deliberation.

However, such considerations of how our species relates to other species pale in comparison to the ELSI matters that arise with information from the HGP regarding *intraspecific* variation and diversity—that is to say, how different humans and groups of humans relate to each other, individually and collectively. The ELSI repercussions of genetic study for health care provision, employment, and insurance are widely discussed, and I do not contend that this focus is off target, for surely these are the predominant ethical, legal, and social questions that concern legislators, businesses, and physicians, with good reason. At the same time, however, I would argue that the HGP has many ELSI consequences that extend well beyond Huntington's disorder, sickle-cell anemia, and breast cancer. Implications of the HGP for human populations and ancestry create thorny issues that relate to social mores and laws. There is no disputing that ours is and has been a tribal species, formed of nomadic or settled clans based around extended family groups. Whether xenophobia is predominantly culturally based or due to inheritance, it seems clear this behavior has long been part of human history, perhaps with good reason. But now that we "know better," thanks to enlightened social norms and attitudes, and thanks to scientific data mines such as the HGP, how do we respond? As we live in a world that rapidly grows smaller, with people traveling to and mating with fellow humans from what were once faraway lands, do we abandon such outlooks as outmoded? How much Native American ancestry must one demonstrate to qualify for special treatment, such as filling a job quota? What if the Hutu or the Tutsi claim Rwanda belongs to them—which group was there first? Should we even have check-off boxes on government forms indicating to which race one belongs?

You may recall the infamous case of Philip and Paul Malone, twin brothers who in 1975 aspired to become Boston firefighters but were not hired because of low scores on the civil service exam.[23] Two years later the Boston Fire Department changed its hiring practices following a court order that mandated preferences for African-American and other minority candidates. The Malone twins, who had previously self-identified as White, and who were white-skinned and fair-haired, retook the test and earned the same scores, but this time were hired after identifying themselves as Black. They worked as firefighters for ten years before a 1988 promotion board noted their racial self-classification. The commissioner who reviewed their applications for promotion knew them personally and was surprised to see both been hired as African-Americans. When confronted, the twins claimed that between 1975 and 1977 their mother told them their great-grandmother was a light-skinned black woman. The department rejected their claim, but not before the story had attracted great renown.[24] The Malones were fired for "racial fraud," and their case, one of the earliest and most publicized episodes of the Affirmative Action era, prompted not only a national outcry over "racial passing," reverse discrimination, and job quotas, but internal investigations in Boston and other cities into dozens of cases of suspected fraudulent racial identification by city employees. Black and Hispanic leaders blasted the Boston Fire Department; one charged that as many as sixty firefighters had gotten jobs through fraudulent self-identification.[25] In more recent years there has of course been a similarly contentious and well-publicized battle over the role of affirmative action in college and university admissions.

I tell this story not to make a point about racial fraud or affirmative action, but to reveal underlying social issues of how we deal with information relating to human history and relationships. What if the Malones were telling the truth? Suppose they really have a great-grandmother of African descent. What does that mean? The truth, at least as far as current science can be said to have a handle on the truth, is that somewhere back down the line, as anthropologist Pat Shipman observes,[26] we are all Africans. The real issue is, how do we deal with this information? In Australia, where aboriginal peoples are accorded special privileges as part of a national reparations and reconciliation program, it is said that virtually everyone has either asked to qualify or at the very least considered qualifying for such special treatment. As an Australian colleague told me, given all the interbreeding of any family that has been in Australia for more than a few generations, anyone can honestly claim to have some aboriginal background. This of course harks back a century to the days of Jim Crow, when the so-called "one-drop" rule held that a person with even a tiny portion of non-White ancestry ("one drop of non-White blood") should be classified as "colored," especially for the purpose of laws forbidding interracial marriage. Some states later adopted the $1/32$ rule whereby you were deemed Black if you had one Black great-great-great-grandparent. As Walter Williams notes,[27] genealogical and DNA evidence might be the only way that White-looking people will be able to exonerate themselves if charged with racial fraud. I don't intend to sink into the quagmire of familiar racial arguments here, but the point is, how "Black" must one be to qualify as Black? Does it matter? And if so, why does it matter? What if the Malones were $1/128$ "Black"?

But wait: Doesn't this conflict with what society tells us everyday? Can't we rather easily tell races apart? Don't we all know what a Black person looks like, or a White person? What is more important to us, what's in our genes or what's on our faces? Or is it, upon further review, more accurate and even more useful to say that we are simply all human? How should we react when assured by many researchers that there is absolutely no scientific basis for classifying humans into racial groups, yet at the same time physicians point out differences in medical conditions and treatments among those very groups? Which group should we believe? What about misleading statistics, like the oft-cited claim that there is less genetic variation between ethnic groups than within them? I will leave it to those far better qualified than I am to dwell more deeply on matters of race, but allow me to emphasize, as I frequently yet gently remind people who inquire as to whether we are indeed related to apes, that the real issue is of closeness of relationship. Indeed, we are related to the trees and grass and bugs outside—to all living organisms, as our DNA demonstrates. We are related to all people. The question, to the extent that it matters, is, how closely are we related? This is where ELSI ramifications pile up in a hurry.

I began by noting that while many of the benefits of the HGP apply to the busy, bustling world of health care, in which we are all entangled almost daily, other benefits relate, at least for the present time, solely and purely to science. Scientists are of course often criticized for not thinking ahead and considering the potential impact of their work: They do it simply because it can be done. Splitting the atom unleashed new energy sources, but at the cost of nuclear weapons. Transgenically modified foods can feed starving people, but with unintended risks of decreasing genetic variability of crop and livestock organisms, triggering allergies in humans, and the potential for gene transfer to wild species. Yet science is a process intended to provide a better understanding of our world, not guidelines for how to live in it. It is society at large that decides what to do with this new information. Knowledge is a tool; it can be used for good or ill.

Is the knowledge that one has a predisposition to colon cancer or heart disease good or bad? It depends, as we all know, on who is asking (an insurer? an employer?) and for what reasons. The same can be said for knowledge about larger groups of people. How do two warring tribes or ethnic groups respond when told they are closely related—do they take this as good news or bad? We would do well to remember that state-sponsored eugenic programs, rightly considered one of the great embarrassments of the twentieth century, flourished not only in Nazi Germany but all over the United States. Once Mendel's ideas about inheritance were disseminated, people thought they could use this information for the betterment of society. Will we do the same in the near future with genetic "enhancement," or will we heed ethicists who bid us think twice before stepping onto another slippery slope? I do not mean to sound a negative, cautionary tone and imply that all knowledge to emerge from the HGP is potentially dangerous or incriminating, or that it should be sealed, like a powerful genie, safe inside an airtight bottle. On the contrary, much of the information we gain from genomic study of humanity should be edifying and ennobling.

As an evolutionary biologist I like to envision a halcyon future in which Darwin's views are as noncontroversial as Galileo's. Just as with the heliocentric solar system, this won't mean we are any less special, only that we know more details about what we are and where we came from. The cynic in me doubts this will happen during my lifetime, but if it does come to pass it will undoubtedly be due in large measure to the HGP and related genetic studies, which should increasingly come to dominate the news and infiltrate society. Many people scoff at scientists yet listen to their physicians. As evolutionary analyses yield not merely new drugs and treatments, but also answers to major philosophical questions, maybe more people will pay more attention.

Notes

1. J. C. Venter, K. Remington, J. F. Heidelberg, A. L. Halpern, D. Rusch, J. A. Eisen, D. Wu, I. Paulsen, K. E. Nelson, W. Nelson, D. E. Fouts, S. Levy, A. H. Knap, M. W. Lomas, K. Nealson, O. White, J. Peterson, J. Hoffman, R. Parsons, H. Baden-Tillson, C. Pfannkoch, Y. Rogers, and H. O. Smith, "Environmental genome shotgun sequencing of the Sargasso Sea," *Science* 304, no. 5667 (2004): 66–74.
2. S. Wells, *Deep Ancestry: Inside the Genographic Project* (Washington, D.C.: National Geographic Society, 2006).
3. L. L. Cavalli-Sforza, *Genes, People, and Languages* (Berkeley: University of California Press, 2001), at 160.
4. B. F. Voight, S. Kudaravalli, X. Wen, and J. K. Pritchard, "A map of recent positive selection in the human genome," *Public Library of Science* 4, no. 3 (2006), e72.
5. M. A. Jobling and C. Tyler-Smith, "Fathers and sons: the Y chromosome and human evolution," *Trends in Genetics* 11 (2005), 449–456; S. Wells, *The Journey of Man: A Genetic Odyssey* (New York: Random House, 2004), 42–44.
6. Wells, *Deep Ancestry*, at 88.
7. S. Olson, *Mapping Human History: Genes, Race, and Our Common Origin* (New York: Mariner, 2003).
8. C. S. L. Lai, S. E. Fisher, J. A. Hurst, F. Vargha-Kadem, and A. P. Monaco, "A forkhead-domain gene is mutated in a severe speech and language disorder," *Nature* 413 (2001): 519–523; W. Enard, M. Przeworski, S. E. Fisher, C. S. L. Lai, V. Wiebe, T. Kitano, A. P. Monaco, and S. Pääbo, "Molecular evolution of FOXP2, a gene involved in speech and language," *Nature* 418 (2002): 869–872; Y. Q. Wang and B. Su, "Molecular evolution of microcephalin, a gene determining human brain size," *Human Molecular Genetics* 13, no. 11 (2004): 1131–1137.
9. R. Lewin and R. Foley, *Principles of Human Evolution*, 2nd ed. (London: Blackwell, 2004).
10. G. Moulson, "Neanderthal genome project launches: Two-year project seeks to decipher Neanderthals' genetic code," *Associated Press*, July 20, 2006.
11. N. Wade, "Scientists plan to rebuild Neanderthal genome," *New York Times*, July 20, 2006.
12. J. Diamond, *The Rise and Fall of the Third Chimpanzee: Evolution and Human Life* (London: Vintage, 2004), at 15.
13. N. Patterson, D. J. Richter, S. Gnerre, E. S. Lander, and D. Reich, "Genetic evidence for complex speciation of humans and chimpanzees," *Nature* 441 (2006):315–321.
14. *Ibid.*

15. M. Crenson, "DNA analysis shows it was messy when humans and chimps parted ways," *Associated Press*, 18 May 2006.

16. Patterson et al., "Genetic evidence for complex speciation"; G. Cook, "Humans, chimps may have bred after split," *Boston Globe*, 18 May 2006.

17. Crenson, "DNA analysis shows it was messy."

18. Patterson et al., "Genetic evidence for complex speciation."

19. Cook, "Humans, chimps may have bred."

20. D. Perlman,. "When humans, chimps were kissin' cousins: Genetic analysis shakes up ideas on when, how the two species diverged," *San Francisco Chronicle*, May 18, 2006.

21. S. Kruglinski, "Are we all Asians?," *Discover Magazine*, May 2006:12–13.

22. Patterson et al., "Genetic evidence for complex speciation."

23. P. Hernandez, "Firemen who claimed to be black lose appeal," *Boston Globe*, July 26, 1989.

24. *Malone v. Haley*. 1989. Supreme Judicial Court for Suffolk County, No. 88-339, July 25, 1989.

25. Hernandez, "Firemen who claimed to be black."

26. P. Shipman, "We are all Africans," *American Scientist*, 91, no. 6 (2003):496–498.

27. W. Williams, "Race board needed," syndicated column, August 6, 2003.

Teaching "Ethical Futures: Implications of the Genome Project"

Tom Segady
STEPHEN F. AUSTIN STATE UNIVERSITY

The teaching of a class on bioethics, specifically the ethical problems presented by the implications of the Human Genome Project, presents several pedagogical challenges. First among these, perhaps, is the dynamic nature of the topic. Second, this is coupled with the assumption the students bring to the class (and this may be shared at least in part by the instructor) that the subject matter of a class has fixed parameters, and that the central issues will not change during the course of the semester. Third, students come from a wide range of academic backgrounds, with majors as diverse as history, biology, physics, communications, and English. As a result, the course must be designed to address a broad spectrum of interests and levels of preparation, and to facilitate the students' awareness of their role as active contributors, regardless of these factors. In order to generate a learning experience that is integrative yet allows the students to develop individual perspectives, a collective project must be designed that allows this to occur. This also demands flexibility on the part of the instructor; he or she may be required, at regular intervals, to tie the students' perspectives and unique set of knowledge back to the general themes of the class. Fourth, much has been made of students' lack of historical awareness or of any meaningful concern for the relevance of history.

Far less attention, however, has been given to the students' awareness of the accelerating pace of social change, and by extension to the importance of the future. Given technological changes that have occurred during their lifetimes, this is perhaps both inevitable and understandable. Moreover, these changes may very well come with a general positive evaluation that combines with a sense of inevitability regarding their development. In a sense, Adam Smith's "invisible hand" has taken on, for traditional college-age students, a felt (but very possibly not perceived) existence in many technical spheres of life. Technology simply *is*. This, in turn, may lead to a tendency to become a spectator of change, and a collector of technology, rather than adopting any critical, involved stance.

Thus, the pedagogical task is a daunting one, and there are very few guidelines for teaching this sort of class. This, however, can become a major advantage if the students are made aware (not just informed) at the outset that the course is both unique and in many ways experimental, and that they will become active players in its creation. They must also, at the very outset, know that their academic competencies are relevant and important. The course was designated a "writing intensive" class, and writing projects began during the first few weeks to address specific ethical problems regarding emerging bases of genetic knowledge and the possibility for genetic remedies. The initial writings also made available to the instructor a valuable tool for an assessment of each student's strengths and weaknesses apart from in-class interaction. This is important; the instructor can then make use of the writings in the first class discussions to illustrate points made by specific students, which in turn can build confidence on the part of class members that their contribution will be valued from the outset. Other students, in turn, can begin to see how their knowledge and perspectives will mesh with those of their peers, thus fostering the beginning of a collective enterprise.

Along with this nurturing of confidence, however, there must be an accompanying challenge, and to students the particular challenge of this course was daunting. In lieu of exams, only the weekly writing papers would be assigned, along with the final project, which was the writing of a book, with each member of the class writing a chapter of the book. Apart from providing a few general suggestions regarding topics and outlining the general requirements for format and length, very little specific information was given; the students would need to generate topics relying on their own intellectual resources and developing interests in the topics. As few students had come to the course with a well-defined subject of interest, the intent was in part to instill a sense of looking at the readings to provide cues for possible subjects. This was one technique to generate a new attitude toward the readings, and not to leave the ideas dead on the page, but to bring them into an active area of concern and significance.

The immediate task, however, was one of moving outside what is often termed the "natural attitude" toward the workings of the scientific and social world and the intersection between the two. To illustrate, a natural attitude toward the medical model contains a series of typically unstated dualities: sickness and health; youth and old age; curable and incurable—and, at a later point in the course, students discovered that this becomes a duality that also can be called into question: life and death. To provide a fundamental, accessible background for the subject, the book *Genome* by Ridley was assigned as one of the basic texts. Ridley's foreword is highly instructive at the outset of a course of this type: In clear terms, he demonstrates how quickly the research has developed—from a 1998 report from the scientists comprising the publicly funded Human Genome Project that they had read 10 percent of the human genome, to the announcement on June 16, 2000, that the "rough draft" of the human genome was complete, based on the aggressive methodology provided by Craig Venter. Additionally, Ridley captures the potentials and dangers that lie immediately ahead, along with the ethical dilemmas that must be confronted, which include side effects of genetic engineering, the widening of inequality, and the destruction of cherished beliefs regarding what is essentially "human."

The objective here is to challenge the sense that change is ordered, proceeds at a measured pace, and leaves core values and beliefs intact. The initial response to such information is a general sense of disbelief and denial that comes as an active "will to ignorance." This also can be positive; it may indicate that students are beginning to actively confront the issues, however unpleasant this may be initially. One student stated near the outset of the class: "This may be more like science-fiction than fact; at any rate, it won't occur during my lifetime or my children's lifetime." One way to interpret this statement is that it pushed the future of genetics at a horizon so distant that it may safely be ignored, or at least diminished in importance. Perhaps this belief was even reinforced by the film *Gattaca*, which does set the ethical dilemmas and practical problems at a distant point. However, after viewing the film, students were able to identify several important issues free of instructor interference or lecturing: for example, emerging genetic discrimination, access to genetic enhancement, the heightened social focus on the importance of one's genetic composition, and the potential hazards even within the family structure itself. This in turn brought more textured reactions to Ridley's quote of an anonymous Scottish solder's exclamation after hearing a lecture by William Bateson: "Sir, what ye're telling us is nothing but scientific Calvinism" (Ridley, 1999:54).

In a course that is designed to have students become engaged as full participants, it is important for the students to develop these sorts of insights. An important, accessible reading was assigned early in the class—

"The Concept of Genetic Malady" by B. Gert. This article works effectively to break down the "duality" presented by the standard medical model between sickness or health. A "malady" is a condition, apart from an individual's rational beliefs or desires, that he or she is "at risk of suffering a harm or an evil—namely death, pain, disability, loss or freedom, or loss of pleasure—and there is no sustaining cause of that condition that is distinct from the person" (Gert, 1996:147). With increasing knowledge of an individual's genome, *being at risk* considerably expands the notion of who has (or will have) a malady: "new genetic discoveries will increase both the number of persons known to have conditions already regarded as maladies and the number of conditions many will come to regard as maladies" (Gert, 1996:147). Students were consistently reminded by Ridley in *Genome* that "genes do not cause disease." However, increasing genetic knowledge does make inevitable the identification of maladies and those who "are at risk" for having them. The students quickly replaced the concept of "malady" for the concept of "illness," and thus both deepened their awareness of the consequences of increasing genetic knowledge and applications. They also realized that, as societal awareness of maladies increase, two ethical issues would come to the fore: Which maladies would be identified as being treatable and which were not, and who would decide this question?

At this point in the class, it became possible to begin considering the contemporary effects of eugenics. This consideration would include, particularly, "back-door" eugenics. Eugenics in this form refers to the lack of a central agency determining these questions, but a plurality of entities (such as insurance companies and employers) that are able to "guide" individual choices through provision of a system of rewards and punishments. This in turn led to several issues regarding privacy, which perhaps heretofore were not seen as significant for a generation that has not faced (to date) threats to individual liberties such as the McCarthy investigations of the past. Outside experts were invited to discuss with the class (and with the university as a whole) emerging privacy issues, and the nexus between privacy, genetics, and legal regulations. Chief among these experts was Professor Albert (Buzz) Scherr, from Franklin-Pierce Law School, whose topic was: "Do You Know Where Your DNA Is? Genetics, Privacy, and the Constitution."

Eugenics raises fundamental sociological questions involving social control, race, and class. The magnitude of the social control issue was introduced by Michel Foucault's "techniques of normalization." By this, Foucault "means a system of finely gradated and measurable individuals in which individuals can be distributed around a norm—a norm which both organizes and is the result of this *controlled* [italics added] distribution" (Rabinow, 1984:20). Prior to the systematized, ordered application of these techniques, it would be possible to visualize persons with widely varying features—intellectually, psychologically, and behaviorally. A bell curve, in this sense, would reflect this in shape, being low and very broad—the extremes would lie quite far from the center. In social terms, this might be best illustrated in Foucault's *Madness and Civilization*: In earlier times or in less ordered, communal societies, social outliers could be incorporated into the society, and perhaps sheltered or even given a special function—e.g., the shaman, highly eccentric artist, or the so-called "village idiot." As the society advanced in order and control, however, these outliers become more akin to "social anomalies"—which are increasingly subject to the techniques of normalization, which Foucault describes as having three separate stages. In the first stage, objectification,[1] the characteristics of the social outlier are identified and classified into increasingly fine gradations. Set in the context of the genome project, this could mean the "discovery" and classification of maladies, and the associated debate regarding "what to do" about these maladies would become invested in structures of power. These "structures of power," however, often exist far from the direct power of the state, according to Foucault—e.g., corporations, scientists, medical communities, psychiatric communities, or organizations. There is no "conspiracy," in the classic sense, to achieve a certain end among these structures of power. The motivations may be economic, religious, scientific, a genuine desire to serve, or any combination of these. At any rate, the motivations tend to be forgotten or sublimated. In a very real sense, Foucault suggests, this may actually strengthen the process of objectification; discourse and criticism are removed from the primary question of effects.

The second process involves *subjectification*, or the labeling of persons according to their maladies. The person becomes a "subject" that can be further classified, and treated in accordance with the labels he or she has been assigned. It is here that Foucault sees the beginning of *external* control, or the ability of agencies (now including the state) to "normalize" the population, and to treat persons outside the norm as somehow deviant and requiring special attention, monitoring, or even restraint. In the field of genetics, this might mean that persons who are so identified are treated in a certain manner. A woman in the class, for example, argued that it was not so much that she was bipolar but that she had been labeled and treated as bipolar that caused her anxiety regarding her future career as well as personal relationships. Populations of individuals labeled "at risk" genetically may face this sort of anxiety in the future.

Finally, it is at the point of *internalization* that Foucault sees the techniques of normalization as most insidious. The locus of control now becomes *internal*, as the person, no longer simply a "subject" who is labeled from the outside, fully accepts that label as part of his or her identify, perhaps even making that their "master status."[2] The person does not believe or act "as if" the classification is real—it has become real, and beliefs and behaviors come into line with this new, accepted reality. While Foucault discusses this in the context of mental illness, students learn the genetic implications visually, through films such as *Gattaca*, in which all but the central character accept their status as being genetically inferior and creating a self-fulfilling prophecy.

Two important consequences then begin to emerge. First, the wide "bell curve" described earlier begins to narrow: The "techniques of normalization" cause what was once divergent but accepted to become subject to increasing control, perhaps through institutionalization, medication, or psychiatric treatment. Moreover, the person becomes a willing participant in this pressure toward normality, accepting (or even seeking out) the mechanisms that will reduce or eliminate the perceived deviance. Second, persons who continue to see themselves as deviant react by re-creating the "dividing practices" at the individual level, retreating from those who would reject them, and regrounding their identities with those whom Goffman (1986) refers to as the "own" (those who share the label) or the "wise" (those who do not share the label but are accepting of it). As a result, the social order becomes reordered around new lines of inequality. In the class, this led one student (who had two brothers with moderate autism) to write a chapter that argued for retaining as a social value "neural diversity."

To see the ethical dilemmas these developments pose, at both the individual and the social levels, students read Michael Sandel's article "The Case Against Perfection" (2004), together with Alice Wexler's (1996) book *Mapping Fate*.[3] Wexler's book, particularly, allowed students to comprehend that their previous sense of the distant and once seemingly impossible future is descending rapidly into the present. In a complex, textured, and thoroughly readable manner, Wexler describes the changes in family dynamics and personal choice that will be associated with genetic discoveries. Further, she weaves through the account a technical discussion of genetic research related to Huntington's disease, which is accompanied by her personal struggle whether or not to know her own genetically determined future. Writing in his book *The Sociological Imagination* (1959), C. Wright Mills asserts that the true understanding of a sociological question comes with a grasp of the intersection between history and biography. Wexler's book conveyed how quickly history—and thus biography—is changing. Her ability to discuss every detail of the struggle to confront Huntington's stands in marked contrast to the past that her family had kept secret just one generation ago. The class was able to arrange a conference-call meeting with the author, which further brought home the immediacy and magnitude of the issues and ethical dilemmas they would confront. Finally, they were asked to contemplate the poem by Sylvia Junot that concludes Wexler's book. The titles of chapters of the students' book, which are listed in Appendix B, demonstrate their willingness to engage in exploring all that is implied by this poem: "when we pull a blade of grass, / up comes the field."

Appendix A: Course Syllabus

SOCIOLOGY 477(H)
ETHICAL FUTURES:
IMPLICATIONS OF THE GENOME PROJECT
SPRING SEMESTER, 2007
Dr. Tom Segady

Books for the course:

Ridley: *Genome.*
Wexler: *Mapping Fate.*

Journal articles for the course:

Munday: "A World of Their Own: Deafness Is Perfect."
Sandel: "The Case Against Perfection."
Gert: "The Concept of Genetic Malady."
Buchanan: "Eugenics and Its Shadow."
Church: "Genomes for All."
Guttmacher and Collins: "Genomic Medicine: A Primer" (optional).
Green: "I, Clone."
Scheitle: "In God We Trust: Religion and Optimism Toward Biotechnology."
Mackie: "The Law of the Jungle: Moral Alternatives and Principles of Evolution."
Green: "Parental Autonomy and the Obligation Not to Harm One's Child Genetically."
Halbfinger: "Police Dragnets for DNA Tests Draw Criticism."
Kimura: "Sex Differences in the Brain."
Gooding: "Unintended Messages: The Ethics of Teaching Genetic Dilemmas."
Ten: "The Use of Reproductive Technologies."
Paul: "Whose Country is This? Genetics and Race."
Nelson: "Prenatal Diagnosis, Personal Identity, and Disability."
(Note: As this is a quickly changing field, there may be other articles that will be included, or articles that supersede the ones that are listed. In that case, those articles will substitute for the ones listed).

Some relevant web sites:

Genome Project Information: www.ornl.gov./hgmis/home.shtml
Medicine and Genetics: www.ornl.gov.hgmis/medicine/medicine.shtml

Ethical, Legal, and Social Issues: www.ornl.gov/hgmis.elsi.shtml
Genome Image Gallery (downloadable): www.ornl.gov/hgmis/education/images.shtml
Genome: Resources for Students: www.ornl.gov.hgmis.education/students.shtml
Careers in Genomics: www.ornl.gov/hgmis/education/careers/shtml
Department of Energy Joint Genome Institute: www.jgi.doe.gov
NIH National Human Genome Research Institute: www.nhgri.nih.gov

Plan for the class:

The class has been constructed to include only highly motivated students who bring a variety of backgrounds to bear on developing phenomena in the field of genetics that will change the future direction not only of society, but of human development as a whole. As a class we will become a "learning community" engaged in absorbing and debating some of the issues that are daily emerging as a result of scientific breakthroughs and applications. In order for us to understand the issues as fully as possible, each member of the class will become an *active* participant in a course that is taught in a seminar fashion. The ideas, empirical findings, and issues of each reading will for the focus of discussion. During the course of the semester, we should be discussing the readings with increasing sophistication, and developing our own perspectives on the continuing debates.

Note on the principle of "value-freedom": In sociology, the principle of value-freedom in its ideal form allows for the most complete, objective understanding of an aspect of the social world. Our task in sociology is not to *apply* our values as we attempt to understand a question. Rather, we *suspend* our value judgments as we analyze the question. We may, however, form (and reform) our own perspectives as a result of our investigations. In a very real sense, this is the final objective of the class: to reach a point of sophistication that allows each of us to hold well-developed and articulate perspectives, while respecting the perspectives of others while analyzing them.

Readings for the course:

Genome: Introduction (pp. 4–10)
Mapping Fate: "That Disorder: An Introduction" (pp. xi–xxv)
"I, Clone"
"The Concept of Genetic Malady"
"The Case Against Perfection"
"Eugenics and Its Shadow"
"Eugenics and the Ku Klux Klan"
Genome: Chapter 21
Mapping Fate: to p. 65
"A World of Their Own: Deafness Is Perfect"

"Whose Country Is This? Genetics and Race"
"The Race and Disease Fallacy"
"Parental Autonomy and the Obligation Not to Harm One's Child Genetically"
Mapping Fate: to p. 162
"Genomes for All"
"In God We Trust: Religion and Optimism toward Biotechnology"
Genome: Chapters 18, 19
"Know Your DNA"
"The Use of Reproductive Technologies in Selecting the Sexual Orientation, the Race, and the Sex of Children"
Genome: Chapters 10, 11
Mapping Fate: finish book
"Sex Differences in the Brain"
"Unintended Messages: The Ethics of Teaching Genetic Dilemmas"

A Note on the Classroom: F172 will be used as our primary seminar room and our project room. It will be available for us to use as an honors meeting room for speakers, organizing events, and designing special topics. There will be a "special collections" section of the room set aside for cataloging important developments during the semester. Also, we will compile a glossary of terms that each member of the class should understand by the end of the class; this will be kept in dictionary form in the room. Also, the glossary will be the first appendix to the book that will be written by the class.

Class Project

The entire class will participate in the writing of a book. The book will consist of individual chapters, with each chapter being written by a student in the class. Selection of the topic for the paper should be made during the first month of the semester. Abstracts and a draft outline of the paper will be due the week before spring break. The chapter should be approximately twenty pages long. It will include a bibliography and notation that will be standardized using ASA format. The chapters will be presented to the class, as well as to those in the university who are interested in the topic, during the final two weeks of the semester.

Individual papers:

Along with the reading assignments for the class, there will be a one-page paper that will summarize and respond to the position presented in the reading. There will be one paper assigned per week, and the paper can be no longer than one page *maximum* in length. The papers will be graded on a scale of 1 to 10. Any paper that receives a rating of 7 or less may be revised. Two papers may be missed during the semester without penalty. Those who complete all papers for the class will have the lowest two ratings eliminated.

Basis for the final grade:

The final grade will be determined in the following manner:
—Chapter for the book: 40%
—Final presentation of the chapter: 10%
—Individual (weekly) papers: 30%
—Quality of class discussion: 20%

Final note: There may be an opportunity for selected members of the class to participate in the summer conference sponsored by Dartmouth College. Details will be made available as the semester progresses.

Appendix B: The Class "Book": Chapter Titles

- Genetics, Race and Hierarchy
- Genes on the Brain: The Argument for Neurodiversity
- Social Eugenics in the Early 20th Century United States
- Genetics and Film
- Face-Off: Nature vs. Nurture
- The Human Genome: A Physician's Point of View
- Genetic Modification and the Ethics of Childrearing
- Who Owns Your DNA? Patenting the Human Genome
- Religion and Genetics
- Genetic Engineering: Issues from the Perspective of Christianity
- Imagining the Future: Predictions from Science Fiction

Notes

1. Foucault also refers to the process of objectification as "dividing practices." See, for example, Foucault's discussion in his *Discipline and Punish* (New York: Pantheon, 1977).

2. For an interesting discussion of this point, see Erving Goffman's *Stigma: Notes on the Management of Spoiled Identity* (New York: Touchstone, 1986).

3. All readings included in the syllabus (Appendix A) are cited in the References section.

References

Buchanan, Allen, et al. 2002. Eugenics and Its Shadow. In Buchanan et al. (eds.), *From Chance to Choice: Genetics and Justice*, Chapter 2. New York: Cambridge University Press.

Church, George M. 2006. Genomes for All. *Scientific American*, January, pp. 45–54.

Foucault, M. 1965. *Madness and Civilization*, trans. Richard Howard. New York: Random House.

———. 1977. *Discipline and Punish: The Birth of the Prison*. New York: Pantheon.

Gert, B. 1996. The Concept of Genetic Malady. In B. Gert (ed.), *Morality and the New Genetics*, pp. 147–166. Boston: Jones and Bartlett.

Goffman, Erving. 1986. *Stigma: Notes on the Management of Spoiled Identity*. New York: Touchstone.

Gooding, Holly C. 2002. Unintended Messages: The Ethics of Teaching Genetic Dilemmas. *Hastings Center Report* 32,2(March-April):37–39.

Green, Ronald M. 1997. Parental Autonomy and the Obligation Not to Harm One's Child Genetically. *Journal of Law, Medicine, and Ethics* 25,1:5–15.

———. 1999. I, Clone. *Scientific American* (Fall): pp. 80–83.

Guttmacher, Allen E., and Collins, Francis S. 2002. Genomic Medicine: A Primer. *New England Journal of Medicine*. November 7: 62–70.

Halbfinger, David M. 2003. Police Dragnets for DNA Tests Draw Criticism. *New York Times*, January 4, p. 16.

Kimura, Doreen. 2005. Sex Differences in the Brain. *Scientific American*, February, pp. 39–44.

Mackie, J. L. 1978. The Law of the Jungle: Moral Alternatives and Principles of Evolution. *Philosophy* 53,206:455–464.

Mills, C. Wright. [1959] 2004. *The Sociological Imagination*. Oxford: Oxford University Press.

Munday, Liza. 2002. A World of Their Own: Deafness Is Perfect. *Washington Post*, March 31, p. W22.

Nelson, James L. 2000. Prenatal Diagnosis, Personal Identity, and Disability. *Kennedy Institute of Ethics Journal* 10,3:213–228.

Paul, D. B. 1995. Whose Country Is This? Eugenics and Race. In Diane Paul (ed.), *Controlling Human Heredity: 1865 To The Present*, pp. 97–135. Amherst, N.Y.: Humanity Books.

Rabinow, P. (ed.) 1984. *The Foucault Reader*. New York: Pantheon.

Ridley, M. 1999. *Genome: The Autobiography of a Species in 23 Chapters*. New York: HarperCollins.

Sandel, M. J. 2004. The Case Against Perfection. *Atlantic Monthly* April:51–62.

Ten, C. L. 1998. The Use of Reproductive Technologies in Selecting the Sexual Orientation, The Race, and the Sex of Children. *Bioethics* 12,1:45–48.

Scheitle, Christopher P. 2005. In God We Trust: Religion and Optimism Toward Biotechnology. *Social Science Quarterly* 86,4 (December):846–856.

Wexler, Alice. 1996. *Mapping Fate: A Memoir of Family, Risk, and Genetic Research*. Berkeley: University of California Press.

"God, Science, and Designer Genes"

An Interdisciplinary Pedagogy

Spencer Stober and Donna Yarri
ALVERNIA COLLEGE

Sir Francis Bacon said, "Nature is often hidden, sometimes overcome, seldom extinguished" (Eliot, 2001). The Human Genome Project has revealed a great deal about our "hidden nature," and emerging technologies enable us to alter our genetic destiny, but the outcomes are uncertain. Science by its very nature follows every avenue of inquiry, and applications are maximized as opportunities arise. We must reflect on these technologies as they arise if we are to avoid potential risks. This chapter describes an uncommon alliance, the disciplines of science and theology, in dialogue for an interdisciplinary course that encouraged student reflection on these challenging issues.[1] We describe the development and delivery of a course entitled "God, Science, and Designer Genes" with examples of how the worldviews of science and religion can engage students in dialogue to increase their reflection on the future implications of these emerging technologies. This chapter will address course development and delivery, scientific and religious worldviews, the challenges of teaching at a religiously affiliated institution, outcomes, and student feedback.

Course Development

Most of the work for this course came at the beginning, in the attempt to develop a course that was truly interdisciplinary, that would attract student interest, and that would fulfill basic core requirements. In this section, we discuss how the course fit into our curriculum, provide the course description, summarize the course objectives, explain the rationale for readings and design of the course, and provide a brief overview of the assignments.

Ultimately the course was offered as a team-taught, interdisciplinary honors class, integrating the disciplines of Biology and Christian Theology. This 400/500-level evening course was available to undergraduates, as well as graduate students in the master's of liberal studies (MALS) program. Students could take it to fulfill one of the following: Theology elective, Ethics requirement, General Education Biology, or MALS elective. As part of our core curriculum, students must take at least one Theology course, an Ethics course in Theology or Philosophy, and one course in Science. The class would also count as an honors class for honors students.

We wanted to develop a course description that highlighted three factors: the interdisciplinary nature of the course, some of the specific questions we would address, and a variety of concrete issues that would likely inspire student interest. Thus, our course description was as follows:

> Modern genetic science is at the state where we can now control our genetic destinies. This course will address both the science behind this phenomenon, as well as some of the ethical and theological concerns, such as: Are we playing God? What is the relationship between religion and science? Issues such as cloning, stem cell research, gender selection, genetic discrimination, as well as other emerging technologies, will be explored through a variety of teaching methods, including videos, case studies, group activities, readings, and discussions.

We developed interdisciplinary course objectives that combined science and theology/ethics. The course objectives were designed to enable students to understand the science behind genetics, to explore the connections between religion and science, and to engage in critical thinking on specific ethical issues. Thus, we came up with nine course objectives, around which we organized the course, and through which we hoped students would achieve the following: (1) gain an understanding of hereditary mechanisms and current theories; (2) gain a greater appreciation for the value of genetic diversity; (3) gain an understanding of current technologies in the field of genetics; (4) gain an understanding of the process of science; (5) consider bioethical issues throughout the course; (6) explore the interrelationship between religion and science; (7) engage in critical thinking about complex ethical issues; (8) explore interdisciplinary ways of addressing questions of religion and science; and (9) gain an appreciation for an ethical and religious approach to scientific questions.

Readings were selected to address both disciplines, as well as raise questions about some specific bioethical issues in genetic science. We chose *Human Genetics: Concepts and Applications* by Ricki Lewis (2005), a widely used and extremely popular work in genetics. We chose the book

Playing God? Genetic Determinism and Human Freedom (Peters, 2003), which raised many important religious questions. In addition, we had a packet of varied reading materials; some examples follow. Nathaniel Hawthorne's short story "The Birthmark" (Arvin, 1946) caused our students to reflect on the desire for a perfect body. Claudia Kalb's *Newsweek* article (2004) "Brave New Babies" challenges future parents to consider "Girl or Boy? Now You Can Choose. But Should You?" Three of the readings dealt specifically with genetic discrimination. Lisa Geller (2002) in "Current Developments in Genetic Discrimination" thoroughly considered discriminatory practices and offered strategies to avoid genetic discrimination. Liza Mundy (2002) in "A World of Their Own" addressed the issue of parents choosing for a disability, in this case deafness. Adrienne Asch (1999) in "Prenatal Diagnosis and Selective Abortion: A Challenge to Practice and Policy," raised concerns about parents choosing against disabilities. The topics were presented in fourteen weeks in the following order: introduction to the course; introduction to genetic science, and religion and science; inheritance of genes and selective abortion; sex selection; DNA structure and privacy issues; multifactorial traits and choosing for disability; DNA evidence, and genes and behavior; cloning; genetic modification of humans and animals; gene therapy and genetic discrimination; reproductive technologies and stem cell research; and evolution. Thus, at the beginning, we wanted to provide a general framework and then move to concrete issues.

There were four components of the class, each worth 25 percent of the grade: class participation, dialogue journals, research paper/action plan, and a take-home comprehensive essay exam. Class participation, whether in the form of large or small group discussions, was not only encouraged but expected, and the course was taught in such a way as to maximize class participation. Dialogue journals were a weekly assignment, in which students were expected to do the assigned readings (which usually included science and religious components) and provide a response to something in the reading.[2] Since we did have a variety of students in the class, some more science oriented and others more religiously oriented, they had the opportunity to focus their journal reflections in light of their own interests. The purpose was to critically reflect on class materials in a one- to two-page double-spaced essay. Students were expected to do one of the following: challenge and question ideas presented by the authors, fellow students, and/or professors; present additional facts, evidence, or arguments to support ideas in the text or to present an opposing point of view; explain why they agreed or disagreed with ideas; share their own opinions and interpretations; and make connections by relating ideas to their educational, professional, or personal experiences. In class we would usually ask a couple of students each week to read their journals aloud in order to "seed" the discussion.

The research paper/action plan was designed to be a semester-long project in which students focused on a real-life social/ethical topic related to the course, subject to our approval. We provided a list of possible, but not exhaustive, topics, such as: Should parents be able to choose the sex of their child? Should the government have a DNA database on all of its citizens? Should society require parents to undergo genetic screening for traits? Should people be denied insurance if they are diagnosed with a gene that causes a disease but they have not yet developed the disease? Each student's paper was to have three sections: an *introduction* in which he or she identified the issue, provided a technical overview of the issue, and addressed the scientific, legal, ethical, and social implications; an *action plan* in which the student presented a plan to mitigate the issue, including specific action steps, which could take the form of "action research" to seek information to consider alternatives[3]; and finally, *anticipated outcomes*, which provided a summary and future implications.

The final component of the course was a take-home essay exam in which the students had to demonstrate their ability to bring together various features of the course. The first question asked about the nature–nurture debate and how it affected at least two specific issues discussed during the semester; the second asked what role ethics and religion should play in science; and the third was a case study addressing the challenges faced in the application of genetic research on a specific issue.[4]

Course Delivery

In the delivery of the course itself, our interdisciplinary pedagogy focused on four broad issues: fairly dividing the class time between disciplines, setting the stage for the course through the use of a film, utilizing team teaching and role playing, and focusing on avoiding dogmatic teaching and instead developing critical thinking skills. We decided to divide the class time primarily into two sections. For approximately the first hour, the students would hear a lecture on genetic science. We utilized technology in the classroom, particularly the interactive white-board, which enabled us to search online databases, present interactive lectures by way of Powerpoint, and save materials for student access on WebCT and the college server.[5] The purpose of the science lecture was to provide the scientific underpinnings for the class discussion for that day. We usually took a short break at that point, and the latter part of the class time was devoted more specifically to an introduction/discussion of the religious/ethical concepts. We would then initiate discussion on the readings by doing one of the following: handing out a case study, generating questions, or having a couple of students read their journals. We discovered that students were very engaged in the discussion and we frequently ran out of time.

Science fiction works such as *Gattaca*, Mary Shelley's *Frankenstein*, *Creator*, and *Jurassic Park* prove valuable when delivering courses that deal with genetic technologies (Firooznia, 2006). To set the stage for our course, and to provide an introduction to some of the important general issues, we began the second class session by showing the science fiction film *Gattaca* (Niccol, 1997). It dramatically presents a future world in which genetic technologies are considerably advanced and have the ability to largely determine (or at least significantly impact) genetic destinies. The film focuses on two brothers, one of whom was born the "natural" way and the other of whom was genetically modified at birth. We handed out a worksheet on the film, which students were to fill out as they were watching it and which then provided a framework for discussion. Some questions we raised were: In what ways were people's identities determined, and what were the advantages and disadvantages? What were the differences between the "valids" and "invalids," and which is preferable and why? In what ways is the nature/nurture debate carried out in this film, and what does it suggest about the relationship between the two? What does the setting of the film tell you about the world of genetic science, and in what way is the kind of world presented better or worse than our own? We discussed the film at the beginning of the following class, but then made reference to the film throughout the course as it pertained to our current discussion. We found that students really enjoyed the film, and we felt that it nicely laid out some of the tensions and ethical concerns raised by the technology of genetic science.

This interdisciplinary course required knowledge and teaching experience in both theology and science. Many of us with adequate preparation can teach outside of our respective disciplines, but teaching within one's discipline allows for the confidence and depth of knowledge necessary to fully engage students in spontaneous and productive dialogue. Team-teaching is an opportunity for each professor to do more than simply teach from his or her area of expertise. We took this opportunity to engage each other in role play to model the kind of dialogue and critical thinking that is necessary to consider ethical dilemmas. We lectured from our areas of expertise and raised questions relevant to our disciplines. We genuinely engaged in team teaching through our combined presence in the classroom at all times, bouncing ideas back and forth between ourselves and the students, and through role playing. At times we spoke from the perspective of our respective disciplines, and at other times we stood outside of our accustomed roles.

A brief example is presented here to illustrate how we utilized role play in team teaching. One article the students read was "The Case Against Perfection" by Michael J. Sandel (2004). We engaged students and each other in a disciplined discussion of Sandel's question: "What's wrong with designer children, bionic athletes and genetic engineering?" (p. 1). Students were surprised that both the theologian and biologist took a cautious view, but for different reasons. One caution, from a biological perspective regarding Sandel's question, is the potential loss of genetic diversity in our gene pool. The value of genetic diversity can be supported with solid scientific evidence, but this does not mean that the science of biology can make moral or ethical claims as to the value of diversity—that is the theologian's task. Using the metaphor "wearing the hat of one's discipline" allowed us to signal students when we were speaking from our area of expertise as a biologist or theologian. However, when we chose to speak outside of our discipline, we simply stated that "we are now removing our biologist (or theologian) hat."

We believe that academic freedom is an essential component of the classroom environment for professors and students.[6] Balanced treatment of dilemmas and controversial ideas is crucial if we are to avoid indoctrinating students to a particular point of view. Presenting these issues in an interdisciplinary format helped to ensure this. Through our role playing and respect for each other's points of view, we created a classroom climate that was safe for the expression of diverse ideas. We were not trying to impose a particular view on specific issues; in fact, we encouraged students to defer judgment until they considered fully the complexity of these issues. We wanted to encourage students not to simply state or to have a position, but to support it with rational arguments informed by knowledge.

Worldviews

We realized at the outset that we needed to consider how we think science and religion are related, and we wanted our students to consider this as well. Students were required to read "Ways of Relating Science and Religion" in *Religion in an Age of Science* by Ian Barbour (1990) as an introduction to the worldviews of both science and religion. Barbour begins by stating:

> The first major challenge to religion in an age of science is the success of the methods of science. Science seems to provide the only reliable path to knowledge. Many people view science as objective, universal, rational, and based on solid observational evidence. Religion, by contrast, seems to be subjective, parochial, emotional, and based on traditions or authorities that disagree with each other. (p. 3)

Barbour organizes his discussion as to how science and religion can relate around four possible paradigms or relationships: conflict, independence, dialogue, and integration. He makes the case that when the differences in the two disciplines are not adequately respected, conflict results. Barbour cautions that when science and religion work independently to avoid conflict, constructive dialogue with the potential for mutual enrichment may also be inadvertently avoided. Barbour advocates for the integration of the disciplines, but cautions that we must be careful not to allow the domain of science to overshadow that of religion.[7]

The late Stephen Jay Gould, a highly respected evolutionary biologist, in *Rocks of Ages: Science and Religion in the Fullness of Life* (1999), enunciates the principle of NOMA (non-overlapping *magesteria*). In contrast to Barbour's approach, Gould is not seeking a diplomatic solution to scientific and religious differences, but rather wants to assert the independent and equal status for science and religion. Thus, Gould fits in best with Barbour's paradigm of independence. Gould asserts: "Science and religion must ask different, and logically distinct, questions—but their subjects of inquiry are often both identical and maximally meaningful. Science and religion stand watch over different aspects of all our major flashpoints" (p. 110). Gould calls for the independent and equal status of science and religion, with reliance on human reasoning as the disciplines dialogue through shared identical subjects of inquiry.

In spite of the obvious differences between our disciplines, we found ourselves in agreement with Gould regarding how science and religion relate to each other. We attempted to integrate the worldviews of the two disciplines in our course, but we also came to realize that we both believe that science and religion should have equal status when considering ethical dilemmas facing humans as a result of genetic technologies. Fifty-eight percent of our students said that they considered religion and science to be compatible, yet we observed firsthand that ethical dilemmas created by emerging genetic technologies were "flashpoints" for consideration in class. These worldviews were used to set the stage for class discussions that were informed by scientific understanding and dialogue regarding moral values with shared respect to minimize conflict. Rational decision making guided by consistent principles, and not fear of consequences, was our goal, as suggested by Gould.

Teaching at a Religiously Affiliated Institution

Teaching such a course at a religiously affiliated institution presented us with unique challenges, as well as unique opportunities. Alvernia College is a Catholic college in the Franciscan tradition, sponsored (and originated) by the order of the Franciscan Bernardine Sisters. Demographically, our student body is fairly evenly divided between Catholics and Protestants, and as our survey results indicated, 86 percent of our students self-identified as "religious" or "somewhat religious." To be a Franciscan institution is to focus on the values represented by the founder of this tradition, St. Francis of Assisi, who emphasized in his mission and ministry especially a concern for the poor and consideration of the natural world. It is the Catholic nature of the institution, however, that became especially relevant and significant as we developed and ultimately taught our course. We briefly consider here the question of Catholic identity; discuss the Catholic document *Ex Corde Ecclesiae* (1990); and address the practical implications for teaching such a course at a Catholic institution.

A discussion of great importance in the world of Catholic higher education is the issue of what it means to be a Catholic institution, and how to retain one's Catholic identity in an era when many other colleges are distancing themselves from their denominational roots.[8] There are legitimate concerns that being "Catholic" is becoming secondary to being a college or university. The question is often posited in a way as to ask how one can be truly "Catholic" and yet still remain committed to academic freedom, which is a core value in higher education. It also raises the question of how "Catholic" does one's teaching have to be, with regard to presenting and/or promulgating the perspective of the Catholic Church on ethical issues. The Catholic Church does take strong stands particularly on medical ethical issues, so we wanted to be careful to be respectful of the tradition while also allowing for open discussion. It is important, then, to consider the recent Catholic document *Ex Corde Ecclesiae* (1990), which focuses on this very question, and to briefly address another significant writing.

The Idea of a University, written by Cardinal John Henry Newman (1852), addresses the unique tasks of a university. He emphasized that while one must allow religion to influence knowledge, and while memorization of material is necessary to the educational enterprise, he believed that the primary task of a university is what he referred to as "enlargement of the mind"—what we today might refer to as "critical thinking." Newman strongly believed that evidence of a truly great intellect is the ability to combine the resources of knowledge from various disciplines and integrate them in a creative and unique way. Teaching an interdisciplinary course on genetics certainly enabled us to attempt such a task. To encourage students to engage in critical thinking is consonant both with the Catholic tradition and with our profession as teachers in higher education. The more recent Catholic Church document, *Ex Corde Ecclesiae* (1990), is a contemporary attempt to grapple with the issue of the nature and tasks of a university, but in particular, Catholic institutions of higher education.[9] *Ex Corde Ecclesiae* (1990) emphasizes a number of important points with regard to what it means to be a Catholic institution of higher learning, including recognizing the relationship of the college and the Church; the critical juxtaposing of faith and reason; the importance of a quest for truth in light of the reality of objective moral norms; and fidelity to the Christian message. Interestingly, the document mentions quite specifically the challenges presented by science and technology, and the subsequent important task of a search for meaning that will benefit society as a whole. A section from paragraph 46 is especially worth noting here regarding the necessity of dialogue between Christian thought and the modern sciences: "This task requires persons particularly well versed in the individual disciplines and who are at the same time adequately prepared theologically, and who are capable of confronting epistemological questions at the level of the relationship between faith and reason." Thus, the overall sense of this document particularly with regard to science and technology is open dialogue that is theologically informed and theologically sensitive.

In light of our denominational affiliation and the document discussed earlier, there are some practical implications that impacted our co-teaching this interdisciplinary course. First, we proceeded a bit more cautiously than we probably would have if we were teaching at a secular institution. While being vigilant in presenting and/or allowing for the discussion of all sides of an issue, we were careful not to advocate for a position in conflict with the Church's teaching. We tried to be fair and accurate in our presentation of the Church's position, but also indicated how other Christians might disagree. The biologist, while he didn't feel himself constricted in discussions on the science behind genetics, considered it important that in some ways our discussion was more "restricted" because we are not a research institution, and therefore our considerations of issues were more theoretical than practical. For example, while we discussed embryonic stem cell research, we could objectively look at both sides of the arguments, but were not in the possibly compromised position of actually engaging in such advanced cell technology. The theologian felt a need to be more sensitive to the Church's position while addressing the ethical issues because she had a greater responsibility to understand and accurately present Catholic theology. The biologist, on the other hand, was usually in the position of teaching biology and not making value judgments, although he did actively engage in that part of class discussion.

Second, we engaged in overt encouragement for students to express their real views, so that they wouldn't feel constricted because we were at a Catholic institution. We continually emphasized that the classroom was a "safe haven" for the expression of diverse views, even those that were in conflict with official Church teaching. Since the Church emphasizes a search for meaning that includes reflection on different disciplines, we believed we were on firm ground here. Third, while theology and biology are distinct disciplines, we viewed this course as a unique opportunity to truly meld the two in a way that would be consonant with *Ex Corde Ecclesiae* (1990). Fourth, we utilized the teaching method discussed earlier, of "taking off the hats" of our respective disciplines, in order to sometimes play "devil's advocate" and distance ourselves from the sometimes overtly religious nature of the discussion. Finally, while we have primarily thus far focused on the challenges, this course allowed for unique opportunities as well. The primary one is that we felt no need to be apologetic about presenting a religious, particularly the Catholic, perspective precisely because of our setting. Thus, not only did we feel in some ways compelled to present the Church's view, we welcomed the opportunity to do so. In a secular institution, we might have had to proceed more cautiously in this regard. Thus, we were able to avoid the sometimes problematic issue of value neutrality as a legitimate goal of higher education.[10]

To sum up, we believe that it is crucial to be respectful of and sensitive to the reality of teaching at a religiously affiliated institution. We think that we made every effort to balance, and largely succeeded in balancing, academic freedom with the religious nature of our institution. The principal ways that we strove to maintain this balance were through role playing and emphasizing to students the freedom to express their own views. The interdisciplinary nature of this course was essential for a fuller understanding of the issues related to our respective worldviews, and for consideration of realistic approaches to the challenges presented by emerging genetic technologies.

Outcomes

Efforts to measure outcomes are rapidly taking hold in teaching. Delivery of course content and the development of skills are easy to measure with traditional assessment tools such as a written examination. However, we wanted to do more than simply measure skills and content knowledge; we also wanted to determine whether and how student attitudes might have changed as a result of this course. We believe student attitudes are measured infrequently because assessment tools for this purpose are difficult to develop. We created a pre- and postsurvey instrument as an attempt to identify changes in student attitudes regarding emerging genetic technologies. In this section we briefly discuss the methods, questions, and outcomes.

This was a small nonscientific sample of twenty-four students who registered for the course and therefore the results are not intended to be representative of the larger population. Participation was voluntary and individual responses were, and remain, confidential (all students participated).[11] Self-reported demographic data indicated that about 48 percent of the students in this class describe themselves as religious, 38 percent as somewhat religious, and the remaining 14 percent as not religious. The female to male ratio was 18:6 and the range of credits earned to date was 33 to 170 credits. The survey contained 57 questions and was administered at the beginning and end of the course.[12] For example, consider this survey question: "Do you want to know in advance the likelihood that you will get a disease that has a cure? Select: yes, not sure, or no."[13] We converted qualitative responses to numerical values for analysis using SPSS and Microsoft Excel (a "yes" was coded as 3, "not sure" as 2, and "no" as 1). The average pre and post response for each question was calculated and then compared using Student's *t*-test.

A simple mathematical analysis was used to determine whether, when student attitudes changed during this course, the students as a class were more likely to "accept" or "reject" the potential ramifications of these emerging technologies. To answer this question, the overall change in each paired pre and post question was calculated by subtracting the average response for each question at the end of the course from the average response at the beginning of the course. For example, an average value of 2.5 on a question at the beginning of the course and an average value of 1.8 on the same question at the end of the course yields –0.7 (the negative value indicates a movement toward a likely to "reject" position). On the other hand, an average value of 1.3 at the beginning of the course and an average value of 2.1 on the same question at the end of the course yields +0.8 (the positive value indicates a movement toward a likely to "accept" position). Eight-four percent of the questions demonstrated some change in value from the beginning of the course to the end of the course (11 changes were statistically significant, Tables 1–5). The change in value for each question was then tallied to yield a small overall negative directional change indicative of likely to "reject" (Table 5). Thus, by the end of this course, our students changed their minds on the vast majority of the survey questions (with several changes statistically significant), and while the overall direction of their average responses was not large, this observation is evidence that our course caused students to reflect on their views regarding potential outcomes of emerging genetic technologies.

We next briefly discuss some important observations that we discovered when we analyzed the survey, as well as offer a rationale for the observed changes. For this purpose, we considered four main themes in the survey. The first theme is *the desire to know one's genetic destiny.* Students did want to know in advance the likelihood that they would get a disease if a cure exists, but they were not sure if they wanted to know if a cure does not exist (Table 1, questions 1 and 2). If knowing one's genetic predisposition to a disease enabled prevention or mitigation, then the decision to diagnose is a distinct advantage and is easy to make. When our students learned that in some cases (such as Huntington's disease) genetic tests are available to prediagnose diseases that have no known cure, a great deal of ambivalence was observed in the class discussion. When the desire to know is considered along with the privacy of one's genetic information, the average responses to the survey questions did not change significantly during the course (Table 2).

The second theme is *the desire to control one's genetic destiny.* Students became significantly less willing (if the technology was available) to have their genes manipulated to extend their life expectancy (Table 1, question 4), and they became significantly less likely to express a desire to live a century or more (Table 5, question 7). This observed significant change may be due to the biological and theological perspectives offered on aging. The biologist's perspective was that senescence may be genetically programmed in organisms. Thus, parents do not compete with their offspring for resources, and the genetic deck is shuffled with each new generation to yield variations that may increase chances for survival in a changing environment. The theologian presented the traditional doctrines of Christian theology regarding "The Fall" and the belief in an afterlife to demonstrate that aging is now a necessary part of human existence.[14]

Table 1

Summary of Responses Regarding Desire to Know and Control Genetic Destiny

Questions are followed by the average student responses: "yes" (coded as 3), "not sure" (coded as 2), and "no" (coded as 1), and the change in value calculated (Δ). A positive change in value suggests that the student has become more likely to want to know and control genetic destiny for given circumstances, and a negative change in value suggests that the student has become more likely to want not to know or control genetic destiny. A *t*-test, paired two-sample for means (two-tailed), was calculated (significance noted with an asterisk).

Question	*pre*	*post*	*Δ*
1. Do you want to know in advance the likelihood that you will get a disease that has a cure?	2.3	2.5	0.2
2. Do you want to know in advance the likelihood that you will get a disease that has no cure?	1.7	1.6	−0.1
3. Do you want to know in advance your life expectancy?	1.3	1.5	0.2
4. Would you be willing to have your genes manipulated to extend your life expectancy? $^*t(23) = 2.563, p = 0.017$	1.9	1.6	−0.3*
5. Do you want to know in advance the sex of your child?	1.7	1.9	0.2
6. Would you want to control in advance the sex of your child?	1.3	1.3	0.0
7. Would you want to know the likelihood that your child will get a disease that has a cure?	2.1	2.4	0.3
8. Would you want to know the likelihood that your child will get a disease that has no cure?	1.9	1.9	0.0
9. If it becomes possible, would you modify genes to control your child's athletic ability?	1.3	1.1	−0.2
10. If it becomes possible, would you modify genes to control your child's intelligence?	1.5	1.3	−0.2
11. If it becomes possible, would you modify genes to control your child's physical attractiveness?	1.5	1.3	−0.2
12. If it becomes possible, would you modify genes to control your child's behavioral traits? $^*t(23) = 2.598, p = 0.016$	1.7	1.4	−0.3*
			ΣΔ = −0.4

Table 2
Summary of Responses Regarding Desire to Know and the Privacy of Genetic Information

Average responses based on "yes, even if I do not want to know" (coded as 4), yes, if I agree" (coded as 3), "not sure" (coded as 2), and "no" (coded as 1)—conditional "yes" answers varied to fit questions. A positive change in value suggests that the student has become more likely to "want to know" with a lower expectation for privacy for given circumstances, and a negative change in value suggests that the student has become more likely to "not want to know" with a higher expectation for privacy. A *t*-test, paired two-sample for means (two-tailed), was calculated—no significant changes on this table.

Question	*pre*	*post*	*Δ*
1. A medical doctor receives genetic information indicating that you will develop a genetic disease in the future with no cure. Should the doctor inform you?	2.9	3.0	0.1
2. A medical doctor receives genetic information indicating that you will develop a genetic disease in the future with a known cure. Should the doctor inform you?	3.0	3.1	0.1
3. A medical doctor receives genetic information indicating that you will develop a genetic disease in the future with no cure. Should the doctor inform your insurance company?	1.7	1.4	−0.3
4. A medical doctor receives genetic information indicating that you will develop a genetic disease with a known cure. Should the doctor inform your employer?	1.3	1.2	−0.1
5. A medical doctor receives genetic information indicating that a patient will develop a genetic disease in the future with no cure and the disease can be passed to children. If the patient has children, should the doctor inform the patient's adult children?	3.2	3.0	−0.2
6. A medical doctor receives genetic information indicating that a patient will develop a genetic disease in the future with a known cure and the disease can be passed to children. If the patient has children, should the doctor inform the patient's adult children?	3.1	3.3	0.2
7. A medical doctor receives genetic information indicating that a patient will develop a genetic disease with no cure and the disease can cause workplace accidents as it progresses. Should the doctor inform the patient's employer?	2.5	2.5	0.0
8. A medical doctor receives genetic information indicating that a patient will develop a genetic disease with a known cure and the disease can cause workplace accidents as it progresses. Should the doctor inform the patient's employer?	2.6	2.3	−0.3
			ΣΔ = −0.5

The third theme is *the willingness to share one's genetic information.* Students were consistently willing to share their genetic information with parents, siblings and a spouse (Table 3, questions 1, 3, and 4), but they became significantly less likely to share their DNA with a cousin, uncle/aunt, close friend, government, and law enforcement (Table 3, questions 2, 5, 6, 10, and 11).[15] This observation may be in part the result of scenes in *Gattaca* where a governmental agency and law enforcement were actively using DNA information to track the "invalids" (those with inadequate genetic pedigrees).

The fourth theme considers *attitudes regarding the application of emerging technologies.* We explore cloning and genetic enhancement for this discussion of outcomes. Our students remained firm in their support of a ban on the cloning of humans, but it is interesting to note that they

Table 3

Summary of Responses Regarding the Willingness to Share a Sample of DNA

Questions are followed by the average student responses: "yes" (coded as 3), "not sure" (coded as 2), and "no" (coded as 1), with the change in value calculated (Δ). A positive change in value suggests that the student has become more likely to share a sample of DNA, and a negative change in value suggests that the student has become less willing to share a sample of DNA for a given set of circumstances. A *t*-test, paired two-sample for means (two-tailed), was calculated (significance noted with an asterisk).

Question	*pre*	*post*	*Δ*
1. Spouse	2.8	2.8	0.0
2. Close friend	2.3	2.0	−0.3*
*$t(23) = 2.326$, $p = 0.029$			
3. Parent	2.9	2.8	−0.1
4. Sibling	2.7	2.7	0.0
5. Uncle/aunt	2.3	1.9	−0.4*
*$t(23) = 2.099$, $p = 0.047$			
6. Cousin	2.3	1.8	−0.5*
*$t(23) = 2.460$, $p = 0.022$			
7. Doctor	2.3	2.2	−0.1
8. Insurance company	1.3	1.1	−0.2
9. Employer	1.2	1.0	−0.2
10. Government agency	1.3	1.0	−0.3*
*$t(23) = 2.015$, $p = 0.056$			
11. Law enforcement	1.5	1.1	−0.4*
*$t(23) = 2.632$, $p = 0.015$			
12. Military	1.5	1.4	−0.1
			ΣΔ = −2.6

became significantly less likely to support a ban on the cloning of animals, plants, and microbes (Table 4, questions 1–4). This may be in part a result of our role play on the subject of cloning. In addition, students were required to read "A Critical View of the Genetic Engineering of Farm Animals" by Joyce D'Silva (2002) and "The 'Frankenstein Thing': The Moral Impact of Genetic Engineering of Agricultural Animals on Society

Table 4
Summary of Responses Regarding Cloning, the Use of Stem Cells, and Genetic Screening

Questions are followed by the average student responses: "yes" (coded as 3), "not sure" (coded as 2), and "no" (coded as 1), with the change in value calculated (Δ). A positive change in value suggests that the student has become more likely to "accept" these technologies and a negative change in value suggests that the student is has become more likely to "reject" these technologies. A *t*-test, paired two-sample for means (two-tailed), was calculated (significance noted with an asterisk).

Question	*pre*	*post*	*Δ*
1. We should ban the cloning of humans.	2.6	2.5	−0.1
2. We should ban the cloning of animals. *$t(22) = 2.077, p = 0.050$	2.2	2.0	−0.2*
3. We should ban the cloning of plants. *$t(22) = 2.859, p = 0.009$	1.5	1.1	−0.4*
4. We should ban the cloning of microbes. *$t(22) = 3.425, p = 0.002$	1.8	1.3	−0.5*
5. Should stem cell research be banned if human embryos are destroyed in the process?	2.0	2.0	0.0
6. Should stem cell research be banned if human embryos are not destroyed?	1.5	1.3	−0.2
7. Should parents be allowed to choose for a disability?	1.4	1.5	0.1
8. Should parents be allowed to choose against a disability?	1.9	2.2	0.3
9. Do you think that the world would be better off if diseases were eliminated?	1.6	1.5	−0.1
10. Do you think that the world would be better off if disabilities were eliminated?	1.5	1.3	−0.2
11. Do you think that the world would be better off if behavioral traits could be controlled?	1.7	1.6	−0.1
12. Do you think that the world would be better off if we could design our children?	1.0	1.0	0.0
13. Should society require all parents to undergo genetic screening for traits?	1.3	1.2	−0.1
			ΣΔ = −1.5

Table 5
Summary of Responses Regarding Religion/Science, Nature/Nurture, and General Questions

Responses were "yes" (coded as 3), "not sure" (coded as 2), and "no" (coded as 1), unless noted otherwise. A positive change suggests students are more likely to agree and a negative change suggests that students are more likely to disagree with the question. A *t*-test, paired two-sample for means (two-tailed), was calculated (significance noted with an asterisk).

Question	*pre*	*post*	*Δ*
1. Are religion and science compatible?	2.3	2.4	0.1
2. Are humans "playing God" when they alter the genetic makeup of a human?	2.5	2.4	–0.1
3. Are humans "playing God" when they alter the genetic makeup of an animal?	2.4	2.2	–0.2
4. Are humans "playing God" when they alter the genetic makeup of a plant?	2.0	1.9	–0.1
5. Are humans "playing God" when they alter the genetic makeup of a microbe?	2.0	1.9	–0.1
6. Should we alter genetic makeup of animals for the benefit of humans? (4 "yes"; 3 "yes, depends on the species"; 2 "not sure"; and 1 "no")	2.3	2.2	–0.1
7. Would you want to live "forever" (a century or more)? $^{*}t(23) = 2.145, p = 0.043$	1.5	1.3	–0.2*
8. Should people be denied insurance if they are diagnosed with a gene that causes a disease but they have not yet developed the disease?	1.0	1.0	0.0
9. How much of what we do is a result of nature (our genes) and how much is a result of how we were raised (nurture/learning)? (5-point scale with number 1 "all nurture")	2.9	3.0	0.1
10. How responsible are we for a behavioral action if it is our genetic tendency? (4 "not responsible"; 3 "somewhat responsible"; 2 "not sure"; and 1 "responsible")	2.1	2.0	–0.1
11. Should employers be allowed to have access to an employee's genetic information? (5 "yes"; 4 "yes, without employee's permission if safety an issue"; 3 "yes, with employee's permission"; 2 "not sure"; 1 "no")	1.9	1.9	0.0
12. What grade do you expect to earn in this course? (A coded as 5, B as 4, C as 3, etc.) $^{*}t(23) = 3.498, p = 0.002$	4.5	4.1	–0.4*
			$\Sigma\Delta = -1.1$

and Future Science" by Bernard Rollin (2002). Students became better informed as to what cloning is and may have therefore been less likely to have an immediately negative reaction to cloning, and more likely to consider the status of humans relative to animals, plants and microbes. Students were surprised to see the ambivalence of both the theologian and biologist on this issue. We shared some agreement, but for different reasons. The theologian raised several concerns, especially those related to the ethical treatment of animals in experimentation.[16] The biologist was resistant to making a value judgment on this matter because his position is that science should not be making value judgments. Our surprise was the observation that students became significantly more ambivalent, which we suspect is a result of our dialogue on the cloning of animals as well as of our own ambivalence.

Student Feedback Regarding the Course

In addition to the course evaluations required for distribution by our institution, we created our own evaluation form in which we sought specific feedback on features of the class we thought were important. We wanted to ascertain what students thought about our course and to determine in particular what features worked best. We also wanted to utilize the feedback for modification of the course if and when we taught it again. There were two types of questions. First, we asked fifteen questions in which students assessed the class utilizing a scale of 1 (poor) to 10 (excellent). Second, we had three open-ended questions. The fifteen scaled questions addressed four different areas: the interdisciplinary and team-taught nature of the course; the use of technology in the classroom; the structure of the class; and the nature of assignments. We were quite pleased to discover, in our analysis of the responses after the course was over, that when averaged, the highest marks were received for "enthusiasm of the teachers" (9.8), "how teachers worked together" (9.4), and "effort to combine science and religion/ethics" (9.0). The responses to the other questions were also generally very positive.

The three open-ended questions included: (1) What did you especially like about the class? What do you think especially worked well? (2) What would you suggest for improving the class? (3) Would you recommend this class to your friends? Why or why not? The answers to these questions also confirmed that we largely achieved what we had hoped to in the teaching of this course. Some specific feedback, especially with regard to the first question, was "Excellent class taught by two well qualified professors. I liked learning about the different perspectives. It especially opened my eyes to scientific methods." Another student commented: "I loved how you split the course with the science topic/ethic topic, and then integrated them for discussion." Another opinion was: "The fact that it was taught by two professors that presented very differing views on controversial topics that arise from the human genome." Additionally, "The teachers worked well together. I liked how one teacher covered half the class and the other teacher finished and they both worked together at times." One student, in the section on improving the class, emphasized, "Do not have this course taught by only one instructor! I think that would be a detriment to the overall material and not be as enjoyable."

Conclusion

Creating and teaching an interdisciplinary course such as we did was both challenging and rewarding work. While biology and theology may in some ways seem to be strange bedfellows, emerging genetic technologies and the reality of the Human Genome Project provide an obvious place for the intersection of these disciplines. While we recognized that one could certainly teach such a course simply as genetic science, or simply as a theological ethics course, we believe that the bringing together of these disparate disciplines provided our students with a deeper understanding of the issues involved. In fact, we would strongly encourage other faculty colleagues in these disciplines to consider engaging in such an endeavor. However, as teachers we also know that we learn as well as teach, and in this course, we both gained a deeper appreciation for the contributions of the disciplines of our faculty colleague. Probably the most difficult thing about designing and teaching this course, and something that we had to be aware of at all stages of the course (including the development of the syllabus itself, especially with regard to the course objectives and assignments), was the divergent worldviews with which we approach reality. While science and religion are not necessarily exclusive in terms of their content and approach, it was a challenge to acknowledge both to ourselves and to our students that at times our disagreement on genetic issues resulted from our different worldviews, and that even when we agreed, it was often for different reasons. What was uppermost in our minds throughout all the stages of our course and its development, however, was our strong and deliberate intention to create a truly interdisciplinary learning experience that would encourage students to accept the best that each discipline had to offer, to demonstrate respect for the different disciplines, and to become critical thinkers as they grappled with the serious genetic issues that will surely impact their futures in very concrete ways.

Notes

1. The terms *theology* and *religion* apply to related but separate areas of inquiry. "Religious studies" explores the phenomenon of religion from a more "objective" perspective, whereas theology explores it from a more "subjective" perspective (from the viewpoint of an "insider"). These terms are used interchangeably in this chapter, although theology was the specific discipline we utilized since we are situated at a Catholic institution.

2. This dialogue journaling technique was adapted by Dr. Nan Hamberger (former Associate Professor of Education, Alvernia College) from a handout prepared by Chris Gordon, St. Cloud State University, St. Cloud, Minn., for the NCTE 1992 Spring Conference, Washington, D.C.

3. In the assigned reading by Lisa Geller (2002, specifically pp. 278–282), Geller provides personal reactions and strategies to avoid genetic discrimination, which may serve as an example for this part of the action plan.

4. The case study used is as follows: *Your friend Jane claims that Mary Shelley's Frankenstein continues to be popular generation after generation because "we just didn't get the message." She convinces you to attend the first meeting of a student group planning to campaign for guidelines that would restrict genetic engineering. But when Jane asks Professor Chromosome to attend, he refuses while grumbling, "Genetic research is not the problem. Application is the problem." You and your friend are surprised at his response because Professor Chromosome's research has enabled doctors to predict Huntington's disease and he is currently working on a gene therapy—surely his research would not be limited! What are the challenges that we face in the application of genetic research?*

5. We received a grant from the Pennsylvania Department of Education for the integration of technology into teaching. This included use of the interactive whiteboard (a large touch-sensitive computer screen on the wall) and WebCT software (a college server supported portal for class management).

6. We recommend that instructors review the "Joint Statement on Rights and Freedoms of Students," pages 261–267 in *Policy Documents & Reports* (commonly called the "The Red Book"), 9th edition, 2001, published by the American Association of Professors (Kreiser, 2001).

7. Barbour maintains that integration actually has three possible forms: "Natural Theology," where arguments for the existence of God rely on human reason; "Theology of Nature," which begins with religious experience and tradition; and "Systematic Synthesis," within which diverse types of experience can be interpreted. Process philosophy (with origins in both scientific and religious thought) may serve as a mediator for this integration. Barbour states his personal view as

supporting the "Theology of Nature" with cautious use of process philosophy (Barbour, 1990, pp. 23–30).

8. A helpful and interesting resource in this regard is chapter 4 of the book *A People Adrift*, by Peter Steinfels (2003). Here he considers the difficulty particularly facing Catholic health care institutions and colleges/universities in trying to maintain their Catholic identity.

9. It is impossible to do justice to the magnitude and impact of this document on Catholic higher education, and thus just the points that we believe are immediately pertinent to our discussion are mentioned. We encourage readers to examine both *Ex Corde Ecclesiae* (1990) and the U.S. Catholic Bishops' (1998) interpretation of it for U.S. Catholic colleges and universities.

10. An interesting article in this regard is "The Dubious Value of Value-Neutrality," by Stephen H. Balch (2006), in which he maintains that the only place for value neutrality in higher education is in the sciences.

11. Requirements of the Alvernia College Institutional Review Board were met.

12. All of the survey questions are displayed in Tables 1 through 5.

13. Unless noted otherwise, a three-point response scale (yes, no, not sure) in contrast to the typical five-point scale was used to force a decision by students.

14. The Christian doctrine of "The Fall," based on the mythological story of Adam and Eve's disobedience to God in the Garden of Eden, resulted in human mortality. The belief in an afterlife, especially heaven, is the Christian doctrine where humans are returned to that original idyllic state.

15. It is interesting to note that this observation did not apply to the military—possibly due to a sense of patriotism by this group of students during the Iraq War.

16. For a fuller discussion of this theologian's perspective, see *The Ethics of Animal Experimentation* (Yarri, 2005).

References

Arvin, N. (Ed.). (1946). *Hawthorne's short stories* (pp. 147–165). New York: Vintage Books.

Asch, A. (1999). Prenatal diagnosis and selective abortion: A challenge to practice and policy. *American Journal of Public Health*, 89(11), 1649–1657.

Balch, S. H. (2006, 16 August). The dubious value of value-neutrality. *Chronicle of Higher Education, 52*(41), B15–B16.

Barbour, I. (1990). *Religion in an age of science,* Volume 1 of the Gifford lectures series. San Francisco: Harper Collins.

D'Silva, J. (2002). A critical view of the genetic engineering of farm animals. In R. Sherlock & J. D. Morrey (Eds.), *Ethical issues in biotechnology* (pp. 261–269). Lanham, Md., and Boulder, Colo.: Littlefield.

Eliot, C. (Ed.). (2001). Of nature in men. In *The Harvard classics: Essays, civil and moral* (Vol. 3, Part 1, of 51). New York: P. F. Collier and Son. Retrieved May 25, 2006, from Bartleby.com, Inc.: http://www.bartleby.com/br/00301.html

Ex Corde Ecclesiae: Apostolic Constitution of the Supreme Pontiff John Paul II on Catholic Universities. (1990, 15 August). Retrieved 25 October 2006, from http://www.vatican.va/holy_fatherpaul_ii/apost_constitutions/documents/hf_jp-ii_a.

Firooznia, F. (2006). Giant ants and walking plants: Using science fiction to teach a writing-intensive, lab-based biology class for non-majors. *Journal of College Science Teaching, 35*(5), 26–31.

Geller, L. N. (2002). Current developments in genetic discrimination. In J. Alper (Ed.), *The double-edged helix: Social implications of genetics in a diverse society* (pp. 267–285). Baltimore, Md.: Johns Hopkins University Press.

Gould, S. J. (1999). *Rocks of ages: Science and religion in the fullness of life.* New York: Bantam Books.

Kalb, C. (2004, January 26). Brave new babies [Parents now have the power to choose the sex of their children. But as technology answers prayers, it also raises some troubling questions.]. *Newsweek,* pp. 44–53.

Kreiser, R. B. (Ed.). (2001). Joint statement on rights and freedoms of students. (2001). In *The Red Book,* 9th Ed. (pp. 261–267). Washington, D.C.: American Association of University Professors.

Lewis, R. (2005). *Human genetics: Concepts and applications,* 6th ed. Boston: McGraw-Hill.

Mundy, L. (2002, March 31). A world of their own [In the eyes of his parents, If Gauvin Hughes McCullough turns out to be deaf, that will be just perfect]. Retrieved April 4, 2004, from *The Washington Post* site: http://www.washingtonpost.com.

Newman, J. H., Cardinal. (1852, Winter 2003/2004). Knowledge viewed in relation to learning: Discourse VI, excerpt from *The Idea of a University. Academic Questions, 17*(1), 7–22.

Niccol, A. (Director). (1997). *Gattaca* [Motion picture]. Columbia: TriStar Pictures.

Peters, T. (2003). *Playing God? Genetic determinism and human freedom,* 2nd ed. New York and London: Routledge.

Rollin, B. (2002). The "Frankenstein thing": The moral impact of genetic engineering of agricultural animals on society and future science. In R. Sherlock & J. D. Morrey (Eds.), *Ethical issues in biotechnology* (pp. 271–286). Lanham, Md., and Boulder, Colo.: Rowman and Littlefield.

Sandel, M. J. (2004, April). The case against perfection. *The Atlantic Monthly, 293*(3), 51–62. Retrieved April 14, 2004, from *The Atlantic* online site: http://www.theatlantic.com/issues/2004/04/sandel.htm.

Steinfels, P. (2003). *A people adrift: The crisis of the Roman Catholic Church in America.* New York and London: Simon and Schuster.

U.S. Catholic Bishops. (1998, 3 December). *Ex Corde Ecclesiae:* An application to the United States. *Catholic News Service.* Retrieved 25 October 2006, from http://www.usccb.org/education/excorde.htm.

Yarri, D. (2005). *The ethics of animal experimentation: A critical analysis and constructive Christian proposal.* New York: Oxford University Press.

Genetics as Religion

Gattaca's Disquieting Vision

Ronald M. Green and Aine Donovan
DARTMOUTH COLLEGE

Gattaca, Andrew Niccol's 1997 film depicting life in a genetic dystopia in the not-distant future, continues to be widely used in courses dealing with ethics or genetics to raise fundamental questions about the uses of genetic information. Less obvious to the naïve viewer, and to scholars who have written on the film or used it in classes, are its pervasive religious themes. In this chapter, we have two purposes: to develop the symbolic and structural elements of the film that have religious significance, and to indicate why it is ethically important to understand the ways in which genetics lends itself to an alliance with religious ideas and ways of thinking.

Gattaca portrays a world in which humans have been reduced to their genes. Human frailty and weakness have been eliminated through genetic engineering. The few parents who opt for a "natural" conception run the risk of producing offspring—"in-valids" or "de-generates"—who will become members of an underclass, relegated to menial jobs with little prospect for advancement or opportunity. *Gattaca*'s world is not altogether dystopian: It is clean and efficient with little crime and abundant order, but it doesn't square with the human impulse toward initiative, perseverance, and self-determination.

One of the fundamental issues at play in the development of genetic technology is the fear of tinkering with a divine order.[1] The film's epigraph begins with a quote from Ecclesiastes (7:13), "Consider God's handiwork, for who can make straight what he hath made crooked," and follows this with a modernist rejoinder by psychiatrist–bioethicist Willard Gaylin: "I not only think that we will tamper with Mother Nature. I think God wishes us to." From the outset, therefore, we are put on notice that *Gattaca* intends to touch on religious themes. Less obvious is *Gattaca*'s persistent attention to the relationship between religion and human genetics. *Gattaca* offers a troubling vision of modern genetics as a form of surrogate religion. In its entirety, the film forces us to confront the consequences of a world where science has appropriated religion.

Several authors have already signaled the idea that genetics plays a quasi-religious role in modern society. According to Nelkin and Lindee, geneticists have described DNA as "the Bible, or the Book of Man, and the Holy Grail . . . a sacred text that can explain the natural and moral order."[2] The idea that our genome is synonymous with our humanness is, according to bioethicist Alex Mauron, gaining strength. The genome as the core of human nature, determining individuality and species identity, has become a kind of genomic metaphysics.[3] This outlook replaces the notion that the *soul* shaped matter into form and imbued an organism with individual characteristics. The sanctification of DNA and the religious notion of the soul as the animating force in humanity share similar theological narratives: "the belief that the things of this world (the body) do not matter, while the soul (DNA) lasts forever."[4] DNA, according to Nelkin and Lindee, is the modern equivalent of the sacred texts of revealed religion—it explains our place in the world: our history, our social relationships, our behavior, our morality, and our fate.[5]

This modern equation of the essential and most important elements of the human being with his or her genetic inheritance is the very premise of *Gattaca*'s dystopian society. As the film begins, we learn that genetic testing and selection have replaced natural conception. Past infirmities have been eliminated and only the best embryos are selected for implantation. The film's hero, Vincent Freeman, results from one of the few natural conceptions or "faith births" occurring outside this screening process, and he pays for this with a life of discrimination and exclusion. Predicted within five minutes of his birth to have a 99 percent likelihood of early cardiac failure and a propensity to violence, Vincent is emotionally rejected by his father, Anton, who refuses to give the child his own name, reserving it for the second-born, genetically selected brother.

From infancy, Vincent experiences discrimination and exclusion. Schools close their doors to him, and his childhood dream of becoming an astronaut is a subject of ridicule. Ultimately, Vincent overcomes all of these obstacles. Combining his own native abilities with the variety of illicit resources available to "invalids" like himself, he lands a job as navigator-trainee in the space agency *Gattaca*. Principal among these illicit resources is DNA provided by Jerome Eugene Morrow, a genetically superior athlete who chooses to sell his genetic identity to a willing buyer. However, what begins as a purely commercial exchange develops into a true "blood brother" relationship. Vincent's most formidable obstacle becomes his own biological brother, Anton, who, in the role of a police detective believes that the long-disappeared Vincent has somehow criminally inserted himself into *Gattaca* and murdered a suspicious supervisor. Triumphing over Anton, Vincent succeeds in joining the prestigious Titan exploration team. As Vincent lifts off from earth, he finally achieves the status of first-born son that his defective genes had denied him.

Cinematography

The message of *Gattaca* is carried by more than its story. Director Niccols uses a *film noir* aesthetic throughout the film to indicate both the sinister nature of this dystopian society and its chilling, inhuman qualities. The prevailing hues of the film are a sickly yellow that reminds us of the "pale corpse coloured rubber and yellow-barreled microscopes" described in *Brave New World*'s famous opening paragraph.[6] Gray and black clothing, whether worn by the clone-like perfect workers at *Gattaca* or the swarming "Hoover" detectives, reveals the monotony and sterility of its gene-dominated society. The dimly lit scenes in *Gattaca*'s modernist office spaces reinforce this mood. The only contrast to this bleak environment is the green of the "Detroit Riviera" beach, where Vincent is conceived in the back of his parents' car (a green Buick Riviera). Water reappears, in various shades of green, yellow, gray, and blue, as a setting for some of Vincent's most vital encounters. Life subjected to perfect scientific control, these aspects of the film tell us, can become spiritual death.

Purity

One of the most religiously significant aspects of *Gattaca* is the film's obsession with purity and impurity. As the work of Mary Douglas has taught us, these concepts are central to religion. Although the understanding of what is pure and impure, clean or defiled, holy or profane may differ from one religious culture to another, virtually all traditions dwell on and privilege these distinctions.

They do so, according to Douglas, because these concepts relate to the essential ordering of the world that a religion affords. Each religious culture holds out a vision of how the component parts of the universe found in nature, society, and individual human beings are meant to be arranged. When these component parts stand in proper relation to one another, we experience a state of holiness, perfection, and integrity.

It follows that anything that disrupts this primordial order is unclean, defiling, or impure. As Douglas reminds us, dirt is "matter out of place." It is never a "a unique, isolated event." Instead, "where there is dirt there is system. Dirt is the by-product of a systematic ordering and classification of matter, in so far as ordering involves rejecting inappropriate elements"[7] (p. 35).

These categories apply preeminently to the human body. Its order both reflects and sustains the larger orders around it. As Douglas says,

> The body is a complex structure. The functions of its different parts and their relation afford a source of symbols for other complex structures. We cannot possibly interpret rituals concerning excreta, breast milk, saliva and the rest unless we are prepared to see in the body a symbol of society and to see the powers and dangers credited to social structure reproduce in small on the human body.[8]

This tells us that all the functions of the body figure importantly in the overall scheme of purity and impurity offered by a religious culture. Sexual conduct, eating, elimination, bodily care, and hygiene are all charged with significance. They afford opportunities for the maintenance of purity or for exposure to defilement, for unsettling or maintaining the social and cosmic order.

If we understand that purity and impurity are central categories for religions, what is striking about *Gattaca* is how thoroughly these concepts are taken over and transformed. In *Gattaca*'s world, ancient Hebrew conceptions of cleanliness and uncleanliness and Hindu notions of caste purity have been replaced by good or bad DNA.

We see this in the film's opening scene. As the title and credits scroll by, vastly magnified images of bodily debris, fingernail clippings, hair, flakes of skin, cascade to the ground in slow motion. As the camera moves outward, we realize that we are witnessing the film's protagonist, Vincent Freeman, as he performs his daily ritual of self-purification. Vincent's ablutions clean away his old self, with the telltale sign of his imperfect genes, and prepare the alternate body, blood, urine, and hair from Eugene, which allow him to pass in *Gattaca*'s society.

Vincent's obsession with genetic "cleanliness" is not just a utilitarian necessity, but also a spiritual quest. His goal is to render himself clean in all the senses that all the members of *Gattaca*'s society, including Vincent, understand. This is nowhere more evident than in the scene following his first sexual encounter with Irene. Having dared to risk sharing his betraying genetic material with his lover, as Irene sleeps, Vincent descends to the beach where, wet seaweed in hand, he scrapes his body clean of his offending cells. Like Jews and Muslims before him, who cannot approach the divine while "contaminated" by sexual emissions, Vincent must recleanse himself in order to reclaim his imperiled new identity.

Family

The theme of sterility versus vitality receives some of its richest elaboration in the film's complex visions of the meaning of family. In fact, *Gattaca* offers two competing visions of family: one genetic and deterministic, the other based on love, choice, and commitment. In both visions, religious ideas and motifs play a role.

Although Anton Freeman, Sr., and his wife, Maria, conceive a child in love, they soon appear to succumb to the society's prevailing vision of what a family should be. When Anton withholds his name from the "defective" child, he encourages the fraternal conflicts that ultimately destroy the two boys' relationship. Maria is more conflicted. Her unconditional love for Anton is evident. She struggles to protect him and even resists a geneticist's efforts to improve the next intended child. Nevertheless, whatever feelings the family has for their unfavored son, they are not strong enough to long sustain his presence in their midst. At an early age, Vincent leaves home, breaks off all contact with his family, and vanishes into society's despised genetic underclass.

It is not hard to see that Vincent's family experience represents a genetic reworking of several Biblical narratives. Foremost among these is the Cain and Abel story. Here it is a genetically "perfected" ideal, which each brother feels called to serve. Superficially, Anton is the "favored" one, and Vincent the one who is rejected. But in some respects, the film inverts the Biblical narrative because as the film develops, Anton plays a Cain-like role, begrudging Vincent's accomplishments and too eagerly assuming his brother's defectiveness. With this inversion, however, the film tells us that in this society, genetic characteristics have become means of human self-aggrandizement and sin. In this respect and in a deeper sense, *Gattaca* stays true to the Biblical narrative. The abiding sin of human beings, evidenced in Anton's genetic pride, is to regard ourselves as self-created and as deserving reward on the basis of our self-conceived achievements. In contrast, Vincent, stripped of genetic pride, marshals his given resources for creativity and achievement. Blood, another feature of the Cain and Abel story, also appears. In a key scene in the film, a young Vincent cuts his finger and offers it in a gesture of blood brotherhood to Anton, who, as his father had done in terms of descent, rejects the offering.

If the relationship between Vincent and Anton (and between Vincent and his "blood" family) reveals how a preoccupation with genes can destroy the bonds between people, Vincent's relationship with others also draws on Biblical themes to develop a competing vision of family not based on blood. Jesus' transvaluation of family, to identify as one one's true kin those bound together by bonds of love rather than biology, provides the informing vision ("Who is my mother, and who are my brothers, and pointing to his disciples he said, 'Here are my mother and my brothers! For whoever does the will of my father in heaven is my brother, and sister, and mother,'" Matthew 12:48–50).

We see this most clearly in the relationship between Vincent Freeman and Jerome Eugene Morrow, two young men of similar age and physical appearance, but markedly different genetic identities and personalities. Genetically among the most gifted, Jerome has responded to adversity (second place in a competition) by becoming cynical and embittered. A failed effort at suicide has left him paraplegic. In Jerome's misfortunes, Vincent sees an opportunity for himself. Their relationship thus begins on a purely commercial plane, with Jerome seeking to profit by selling his still valuable genetic material to Vincent, who, as a "borrowed ladder," needs them to fulfill his dreams.

But what began as a commercial endeavor, evolves into something more. When Jerome's drinking, by contaminating the urine samples, imperils Vincent's first interview at the aerospace company *Gattaca*, a confrontation erupts between the business partners that moves their relationship toward brotherhood. Looking at Jerome, Vincent says, "It's not too late to back off. This is the last day you'll be you and I'll be me." Jerome's expression suggests that he accepts this self-transformation, and ultimately their union of spirits. In this moment, both men relinquish the separate identities that had marked them and move toward deep spiritual brotherhood.

The union between Vincent and Jerome reaches its apogee near the film's end, with Biblical religious themes deliberately recalled to mark the extent of the transformations. Name changes have already signaled the union of the two into one. Jerome has abandoned his given name for the genetically significant name of Eugene, while Vincent adopts the name Jerome for the new person who synthesizes the two. Jerome/Vincent has succeeded in winning a place on the Titan mission, and Eugene, who earlier resisted providing an adequate supply of bodily materials, has now filled their freezer with years worth of blood and urine.

As Vincent surveys the provisions that Eugene has prepared for him, Eugene wheels into the room. Vincent, still amazed, asks: "Why have you done all this?"

Eugene responds, "So Jerome will always be here when you need him."

The Last Supper motif of this scene underscores the process of self-emptying that Jerome/Eugene has undergone.[9] By bestowing the gifts of his whole being on Vincent, he has restored himself and spiritually completed the earthly vocation that was only poorly imagined by the human genetic manipulators who created him.

The film ends as the new Jerome boards the Titan-bound spacecraft. He carries a gift from Eugene to be opened only at his ascent. We see a close-up of flames as the rocket engines' igniting fills the screen, and the images shift to the incinerator inside the two men's apartment. There another ball of fire engulfs the unseen figure of Eugene, his terrestrial work accomplished. Around his neck we glimpse the medal he had received for his athletic achievements. Depicting two men swimming side by side, it represents the epitome of his genetically programmed career—and failure. Now it takes on new meaning as a representation of the relationship between himself and Vincent. Aboard the spacecraft, Vincent rises to the heavens as the flames consume the medal and Eugene. He opens the gift from Eugene, a card contain a lock of Eugene's hair. In a voice, Vincent/Jerome says, "For someone who was never meant for this world, I'm suddenly having a hard time leaving it . . . Maybe I'm not leaving. Maybe I'm going home." In this sacrificial event, the brothers have redeemed each other and, in the process have re-created the world. A new communion of love and brotherhood has replaced the false hope of a scientifically created "perfect" community that resulted in a sterile regime of domination, envy, and enmity.

Somewhat antithetical to the prominence given to the theme of mutual support in the Vincent–Eugene relationship are the strong motifs of individualism evidenced in the film. Some viewers of *Gattaca* have argued that beneath its scientific veneer lies a traditional American hero film with the Horatio Alger myth of the triumphant individual. Certainly there are many aspects of the film that support this interpretation. But at the same time, the film interrogates this viewpoint and poses a countervision in which family and community are the essential ingredients of self-fulfillment.

Nowhere is this ambivalence toward individualism better displayed than in the figure of Lamar, the tech assistant at *Gattaca* who verifies each employee's genetic identity. After Vincent has surmounted all the obstacles that have impeded his goal of becoming an astronaut, Lamar conducts a final check and Vincent's true identity is exposed. As Vincent resigns himself to defeat, Lamar explains that his own son had shown himself to be "less than" the genetic designers had promised. Lamar then pushes a button to suppress the information of Vincent's deception, explaining, as he does so, how much his own son admires Vincent's achievements. Although Lamar can be seen as someone motivated by the myth of individual achievement, it is also apparent that Vincent could never succeed without his help. In this respect, Lamar joins Irene and Eugene, who have also lied to protect Vincent, in a small community of supporters that together make Vincent's success possible. All of these people have experienced some measure of genetic discrimination. Their suffering has sensitized them to human vulnerability and inspired compassion. Like the early community of Christians, they are marginalized people drawn together as a family not by ties of blood but by love.

Implications for Teaching

Why is it important for students to consider the pervasive religious themes in a film like *Gattaca*? One answer is that these themes are not easily perceived. Students who focus on *Gattaca*'s obvious social and ethical dimensions—"Should we alter our genes?" "Should we ever permit genetic discrimination?"—are likely to miss the substrate of religious symbols and concepts that permeate the film. Bringing these symbols and concepts to students' attention is thus a worthwhile exercise in hermeneutical sophistication. This affords teachers an opportunity to convey the important insight that works of art should not necessarily be viewed only on one level. This exercise also reinforces the importance of a multidisciplinary approach.

A second reason for highlighting *Gattaca*'s religious themes is to afford teachers of religious studies the opportunity to point up the omnipresence of religion in human cultures. *Gattaca* shows that the concerns that have always inspired the religious imagination persist even in a world of apparently radical secularization and domination by science. Family relationships, the extent of human freedom, the appropriate exercise of human sexuality, our place in the cosmos, and the dividing lines between order and disorder—these questions preoccupy us as human beings. The great religions seek to address them, and their answers remain relevant even in a world where science and technology appear to subordinate religious concerns.

Appendix: Study Guide Questions

The use of film in a class dealing with the issues raised by the Human Genome Project is a wonderful teaching technique. However, the quality of discussion very much depends on preparation on the part of both the instructor and the students. The following list of questions is designed to enrich the educational value of the film and to prepare students for a discussion of the ethical and religious dimensions of *Gattaca*.

1. What is the significance of "names" in the film?
2. What does this film tell us about the nature and risks of genetic discrimination?
3. Forensic DNA plays a large role in the film; what are the messages about that?
4. Does religion play a role in this film? How? In what ways?
5. Describe the aesthetics of this society—architecture, clothing, lighting, interior design, etc.
6. What is the significance of the sea in this film?
7. How has genetic information affected romance in this society? Competition?
8. How is "cleansing" used throughout this film?
9. What does this film say about family and brotherhood?
10. Who is Lamar, and why is he important in the film? How does Lamar's disappointment in his son's genetic potential (or lack of) affect his relationship with Jerome?
11. Gattaca is sometimes seen as a "Horatio Alger" story of one person's triumph over adversity. Is it?
12. How does this film serve as a cautionary tale about the asymptomatic sick?

Notes

1. For discussions of the religious implications of new genetic and reproductive technologies, see David H. Smith and Cynthia B. Cohen, eds., *A Christian Response to the New Genetics: Religious, Ethical, and Social Issues* (Lanham, Md.: Rowman & Littlefield, 2003); Ted Peters, *Playing God: Genetic Determinism and Human Freedom* (New York: Routledge, 2003). Also, Audrey R. Chapman and Mark S. Frankel, eds., *Designing Our Descendants: The Promises and Perils of Genetic Modifications* (Baltimore, Md., and London: Johns Hopkins University Press, 2003).

2. Dorothy Nelkin and M. Susan Lindee, *The DNA Mystique: The Gene as a Cultural Icon* (New York: W. H. Freeman, 1995), p. 41. See, too, Tom Frame, "'Gattaca' and the Challenge to Christian Anthropology," *St. Mark's Review* 174 (Winter 1998), 27–34.

3. Mauron, Alex (2001). "Is the Genome the Secular Equivalent of the Soul?," *Science* 291 (5505, no. 2), 831–832.

4. Nelkin and Lindee, p. 53.

5. Ibid., p. 57.

6. Aldous Huxley, *Brave New World* (New York: Harper & Row, 1932), p. 3.

7. Douglas, Mary, *Purity and Danger: An Analysis of Concepts of Pollution and Taboo* (New York: Praeger, 1966), p. 115.

8. Ibid., p. 115.

9. A similar point is made by Elizabeth Barnes, "Gattaca and A.I.: Artificial Intelligence: Views of Salvation in an Age of Genetic Engineering," *Review and Expositor* 99/1 (2002), 59–70.

Teaching Genomics and Law by Exploring Genetic Predictions of Future Dangerousness

Is There a Blueprint for Violence?

Erica Beecher-Monas
WAYNE STATE UNIVERSITY LAW SCHOOL

As a concrete example of how genomics might inform an important legal issue, I focus, in one law school seminar that I teach, on future dangerousness predictions in the courts. These predictions are generally made at sentencing proceedings by experts, usually psychiatrists, sometimes psychologists, and are based on the idea that the expert can predict whether the defendant is likely to be violent in the future. Surprisingly, although the stakes are high in these predictions, they receive little judicial or legislative scrutiny. For more than twenty years, the American Psychiatric Association has been advising the courts that nothing in the training of these experts would permit them to make such predictions.[1] Although courts and legislatures are well aware of the unscientific nature of these predictions, they nonetheless continue to demand them.

We typically begin the first few class sessions with an exploration of the legal settings in which predictions about violent behavior are made. These include some legal areas with the highest stakes: death penalty sentencing and sexual offender commitments for indefinite periods of time. We then explore the use of expert testimony in these areas, and discuss the problem of judicial gatekeeping failure. We also explore the alternatives to clinical predictions.

Responding to the continued legislative and judicial demand, researchers attempting to improve the accuracy of future dangerousness predictions developed actuarial instruments based on risk factors for violent recidivism. None of these instruments is highly predictive. Whether future dangerousness predictions can meet standards of scientific validity, and what—if anything—can be done to improve them are highly debatable issues. The question posed by behavioral genetics is whether molecular biology can improve this dismal record. Examining the answer to that question is the focus of the balance of the class.

The quest for a utopian world without crime is not new, nor is the idea that crime has a hereditary link.[2] My students come to law school equipped with cultural assumptions. Among these assumptions that "everyone knows" is that the cycle of violence is repeated across generations. The class explores the issues posed by Scandinavian twin and adoption studies, which are widely touted as favoring a genetic role in crime.[3] For this section of the class I typically assign selections from Debra Niehoff, *The Biology of Violence* (1999). Recently, alleles of specific genes, like that transcribing for monoamine oxidase A ("MAOA"), have been identified and linked with propensities to violence.[4] These studies have been cited in criminal cases, and an understanding of human genomics will become increasingly important to law. We also read the Caspi studies and examine their implications in light of various criminal defenses.

While some students initially come to class inclined toward a philosophy of preventive detention, they soon begin to realize that the trouble with basing a utopian vision of a world without crime on genetic propensities is that we have the unfortunate history of eugenics to remind us of the dangers. While we might like to believe that the bad old days are behind us, the shocking absence of scientific scrutiny for eugenics assertions has managed to persist in the astonishing failure of courts and legislatures to examine the scientific validity of expert future dangerousness predictions. In this seminar we explore how information from the biology of violence—including genomics—could improve the way predictions are made, and the importance of making sure that the information is tested, scrutinized, and properly limited, so that the promise of science is not once again perverted into the cynicism of political expediency. One of my goals in teaching this class is to help students to understand that while genes may constrain, influence, or impact behavior, they do so only in concert with each other and the environment—internal and external—of the organism carrying the genes. Moreover, genetic information, no matter how detailed, cannot answer the fundamental legal questions of what responsibility and will mean in our society. These are the crucial issues I wish to explore with my students.

In order to illuminate the dangers of the uncritical use of bad science by law, my students need to learn some legal history. Preventive crime control measures during the eugenics period were based on mostly wrong and mostly unchallenged notions of the connection between behavior and heredity, and manifested primarily in such "preventive" measures as mass sterilizations, indeterminate sentencing, and ethnically biased immigration laws. Tagging along with preventive crime control measures was the idea that the hereditary nature of crime meant diminished responsibility. Thus, at the same time that legislatures were enacting hereditarian-based preventive crime control measures, Clarence Darrow managed to save his clients Nathan Leopold and Richard Loeb from the death penalty by arguing that they were the "product of heredity."[5] These twin themes linking genetics to crime prevention and exoneration have resurfaced, rejuvenated and reinvigorated by the massive attention given to genomics. Indeed, behavioral genetics has been heralded by some as the future of criminal justice. I challenge my students to explore the meaning of those assertions.

The question now facing social policymakers—including the future lawyers and judges whom I teach—is whether and how to use these scientific advances in a manner that avoids the pitfalls of eugenics. While it is true that involuntary sterilization statutes were largely repealed by 1974, determinate sentencing guidelines have imposed some uniformity on punishment, and none of our immigration laws retains language of ethnic bias, the uncritical use of science continues in the criminal justice system's reliance on future dangerousness predictions; apart from Clarence Darrow's clients, however, defendants who have attempted to use the idea of innate propensities to negate individual responsibility have met with great resistance.[6] Unless the accused independently meets the criteria for legal insanity, criminal defenses built on behavioral genetics have been overwhelmingly defeated. For example, as soon as studies linking XYY chromosomal abnormalities with criminal propensities became available, defense attorneys argued that their clients should be exonerated.[7] In the United States, the four cases that attempted such a genetic defense were unsuccessful. Claims that a genetic predisposition to addiction rendered the defendant's actions involuntary have also failed overwhelmingly.[8] Claims for sentencing mitigation have been only slightly more successful.[9] A botched attempt to obtain expert testimony to mitigate the sentence of a capital murder defendant on the basis of MAOA gene abnormality similarly failed.[10]

Recently, however, the Ninth Circuit ruled that it was ineffective assistance of counsel for the defense to fail to present mitigating "psychobiological" evidence at a capital sentencing hearing.[11] This decision raises important questions about what I should be teaching my law students:

future prosecutors, defenders, and judges of people with Landrigan's problems. In *Landrigan v. Stewart*, the petitioner had filed a habeas corpus petition claiming ineffective assistance of counsel, for failing to present mitigating evidence during the penalty phase of his capital murder trial.[12] The petitioner, Jeffrey Landrigan, contended that although he refused to permit his birth mother or ex-wife to testify, he would have cooperated had the attorneys presented expert evidence that his "biological background made him what he is."[13] The available evidence that his counsel could have presented at Landrigan's sentencing hearing, but did not, included the facts that Landrigan had been abandoned by his alcoholic mother at six months, and adopted by alcoholic parents; his biological father was on death row in Arkansas; and Landrigan's substance abuse began at an early age. The *en banc* appeals court noted that, subsequent to Landrigan's conviction and sentencing, an expert had prepared a report pursuant to a thorough neuropsychosocial evaluation, concluding that Landrigan's genetic makeup, in utero exposure to teratogenic substances, early maternal rejection, and troubled interactions with his adoptive family resulted in disordered behavior that was beyond his control and left him unable to function in a society that expects individuals to operate in an organized and adaptive manner.[14] Defense counsel had failed to uncover any of this information, although counsel had the obligation to conduct a thorough background investigation. According to the Ninth Circuit, this is "the very sort of mitigating evidence that might well have influenced the judge's appraisal of Landrigan's moral culpability."[15] None of this information was presented at sentencing.

The dissenting opinion in *Landrigan* agreed with the majority that counsel's assistance was ineffective, but disagreed that the factors would have been sufficiently mitigating, pointing out that (if proved) the allegations of Landrigan's predisposition for violence would have merely emphasized his potential for future dangerousness. Since future dangerousness is the major means of persuading the sentencing jury[16] that a convicted defendant poses a threat to society and thus merits the death penalty, this is not a trivial point. The most hotly debated topic in juror sentencing deliberations—next to the crime itself—is the issue of the defendant's dangerousness on return to society.[17]

Students need to know that jurors focus on future dangerousness whether or not it is mentioned by prosecutors or presented as evidence in the penalty phase of the case, and regardless of the presence or absence of expert testimony.[18] It takes precedence in jury deliberations over any mitigating evidence, such as remorse, mental illness, intelligence, or drug/

alcohol addiction, and any concern about the defendant's behavior in prison.[19] This crucial deliberation, however, is taking place in an intellectual vacuum.

In order to reach a just decision, jurors are supposed to be provided with "information relevant to the imposition of the sentence and provided with standards to guide its use of the information."[20] If the future dangerousness determination is as crucial as the studies suggest, jurors need information about predicting behavior. At capital sentencing hearings, both prosecution and defense present testimony and argument. Expert testimony is prevalent.[21] The experts, however, are not providing this information.[22] Clinical predictions commonly presented at sentencing hearings are speculative at best.[23] Even the most scientific predictions based on thorough examination, diagnosis of mental symptoms, past patterns of behavior, and probabilistic assessment, are wrong nearly as often as they are right.[24]

Death, incarceration, and preventive detention based on predictions that the courts acknowledge as scientifically flimsy seem somewhat cynical, to say the least. If the accuracy of these predictions can be improved with a more thorough understanding of the biology of violence, wouldn't we more closely approach the fundamental requirements of a just system? Although there are a number of empirically based actuarial risk assessment instruments that statistically correlate specific factors in past behavior (a pattern of past violence, for example), their circumstances (poverty, for example), and their attitudes toward others (failure to marry or form equivalent relationship); their medical and psychiatric history (age when problems began, and any injuries to the brain); and substance abuse (alcohol or drugs), even the best risk assessment instruments correlate only moderately with violent recidivism. These instruments still entail the likelihood of false positives—condemning to death or indefinitely incarcerating people who would not commit another violent act. Many of the risk factors measured by actuarial assessments of violence risk may be tied to an underlying biological function. But no testable theoretical basis has been advanced for why the risk factors correlate with violence. Thus, a major problem with each of the risk instruments is its failure to correlate the risk factors with the biology of violence and to articulate an hypothesis for the mechanisms of violence. Until we have an understanding of how violence occurs, we will have little ability to control or predict it.

Having explored the tenuous basis for future dangerousness predictions, and the implications for justice that result, the class is ready to explore the current explosion of knowledge about how biological (including genetic) factors, combined with environmental factors—such as stress, including drug and alcohol abuse—can increase the chances that a particular individual will become violent.[25] Violence is one of those perpetually perplexing human attributes that has fueled speculation about human nature for centuries. Sexual violence is even more puzzling. Thomas Hobbes envisioned humans' natural state as a war of all against all. Evolutionary biologists argue that aggression and violence are our evolutionary heritage, part of our normal repertoire of responses to conflicts of interest in competition for resources. Geneticists, on the other hand, search for rare traits that are heritable and have identified certain genes—like that coding for MAOA—correlated to violent behavior. Social scientists contend that "bloodletting and savagery are . . . unusual . . . [and] shocking." Is there a way to synthesize this information?

Starting from the perspective of evolutionary biology, the class explores the situations eliciting aggression and violence, which commonly involve competition for reproductive resources.[26] In humans, demographically speaking, most violence is perpetrated by young males against other unrelated young males, under circumstances in which risk taking improves their chances of reproductive success. Even spousal murders and infanticide can be viewed as aggressive responses to threats to reproductive success. Most spousal homicides result from male attempts to exert control over women's reproductive capacities and women's efforts at independence from coercion. Whether rape is a reproductive strategy is more contentious, with one camp claiming it as a "natural" genetically based reproductive strategy, and the other camp claiming that rape is not a reproductive, but an aggressive and dominance strategy, a weapon, not an appetite. No one has yet advanced an evolutionary strategy as the basis for pedophilia.

We also discuss the role of behavioral traits, which, as well as behavioral strategies, are involved in violent behavior. The students are well aware that individuals vary in their responses to given situations, and this must be due at least in part to genetic variations among individuals. Differences in personality account for a large percentage of behavioral differences. The class discussion turns to two major factors that inform these differences in aggressive or violent behavior: stress hormones and sex hormones.

The class explores some basic literature on the stress response.[27] Although most violence is perpetrated by young men, violent tendencies can develop prenatally, in early infancy, or can emerge after puberty. Environmental factors often play a role at each stage in this development. For instance, disruption of the early environment can cause increased nervous system sensitivity to stress in animals. Variations in maternal care influence the development of hormonal stress responses, as does social isolation. Depression, violent aggression, and antisocial personality have all been linked to problems with the stress response. Although each of these disorders has a different pattern of expression, all are associated with abnormal endocrine feedback, norepinephrine and serotonin functions, and altered glucocorticoid levels. It may be also be that neuropsychological impairments disrupt normal development and increase vulnerability to poor social environments.

An interesting interaction of nature and nurture is postulated by Tooby and Cosmides, who suggest that early environmental cues may calibrate the stress response irrevocably.[28] These early cues may indicate the kind of social environment the child has been born into. For example, "[v]iolent treatment in childhood increases the likelihood that a person has been born into a social environment where violence is an important aspect of social instrumentality." Thus, exposing a child to violence may permanently lower the threshold for activation of the "fight or flight" response, and may account for the disproportionate aggression in adults who were abused as children. This appears to be the mechanism of MAOA gene activation.[29]

Hormones, which are regulated by the hypothalamus and the autonomic nervous system, can play a role as well. Testosterone affects levels of aggression even in the womb, at least in mice, whose intrauterine positioning between sibling males and females affects levels in postbirth aggression.[30] Testosterone in boys surges at age ten, rising to a plateau by age fourteen, when aggressive behavior starts accelerating. However, delinquent behavior, adjustment problems, and rebelliousness are actually more likely to be associated with lower testosterone levels. Apparently, testosterone, like the stress response, functions as a sensitizer, permitting males to match the social environment with an appropriate response.

Structural dysfunction may also contribute to violent behavior. Damage, decreased uptake of energy-producing glucose, reduced blood flow (or metabolism) to the frontal lobes, and reduced function have all been observed in the frontal cortex of violent individuals and murderers.[31] Reduced volumes of the frontal and temporal cortex have been observed in violent patients.

In sum, I try to get the class to understand that violent behavior is both universal and individual, both normal aggression gone awry and pathological breakdowns in biological systems, both nature and nurture. Many hormones and neurotransmitters, and a multitude of genes, contribute to aggression, along with the social environment, context and timing. Personality and genetics are involved. So are social circumstances, like status, future prospects and relative wealth. Violence is complex behavior.[32] Why wouldn't identifying the genetic contributions aid the accuracy of future dangerousness predictions?

In answering this question, the students should begin to discover that the explosion of scientific information related to sequencing the human genome has created an unfortunate misunderstanding by many, lay and scientist alike, as to the role that genes play in behavior.[33] The assumptions that I encourage my students to question go something like this: Genes made us the way we are, and therefore, we are driven by a set of rigid, genetically determined predilections towards some behaviors. The force of this genetic dictatorship makes us less responsible for our actions, mainly because genetic "determinism" is absolute and irreversible.[34] But the major factor influencing the development of the brain is neither nature nor nurture, not one or the other, but both, in concert, together with random "developmental noise" that leads to great individual variation. In order to help the students in thinking critically about these issues, we read and discuss several books written for general audiences, including Ridley,[35] Lewontin,[36] and Niehoff.[37]

There are a number of ideas we explore through these readings. First, genes do not make behavior, they make proteins. In order to generate behavior, you need a complex nervous system. Various proteins may make a nerve cell, but nerve cells by themselves do not generate behavior. To generate even simple behavior, brain cells must be connected together accurately into working circuits, e.g., motor system, visual system, etc. Within each of these systems, every cell needs to find its specific synaptic targets, other nerve cells with which they will synapse and communicate to create the working nervous system. If that search is thwarted, the cell may never find a target and die. This is the process of "programmed cell death."[38] The process of establishing appropriate connections in many brain regions is essentially stochastic, and, as such, is extremely vulnerable to environmental conditions. For example, if the environment of the developing fetus is disturbed during the time when the main links are being made (first and second trimester), abnormal connections may be made. If, for example, oxygen deprivation causes too few cells to survive, mental retardation may occur. If there is an overexpression of growth factors so that too many cells may survive, it may result in some forms of schizophrenia. Between these extremes there must be a great number of milder effects on cell survival, none genetically determined, that may give rise to individual differences.

Even if all goes well and one group of cells connects appropriately with another, all that creates is a very basic nervous system. Such a nervous system is still incapable of "behavior." Only when all the brain systems are in place, networked together, and exposed to an appropriate environment, is behavior—normal or abnormal—possible. The combination of a good genetic program and the exposure to a proper environment creates a well-functioning nervous system.

All traits, from biological traits like hair and height to complex psychological traits like intelligence, are caused by interdependent interactions of genes and the environment. Thus, the sequence of events is that genes make proteins, that make cells, that make connections, that make systems, that make brains, that will only then cause behavior that is reacting to something in the environment. At every point in this building and tuning process, the environment plays a hand, for better or for worse. Are genes important? Yes. Is the environment important? Yes. Is one more important than the other? No. You have to have genes to make proteins, and you have to have an environment at every step of the way in order to ultimately generate appropriate or inappropriate behavior.

At the conclusion of the class, I hope the students will understand that human beings live in a complex world of interdependent conditions. The easy answer, to find a "genetic" basis for locking a violent individual up and throwing away the key or, alternatively, exonerating that individual from criminal responsibility, is unsupported by the evidence. We know that at least some people are capable of changing their behavior throughout their lives. We know that the end result of what each of us becomes is influenced not only by genes, but by multiple environmental factors at many points during our lifetime, from conception to birth and adulthood.

I hope my students will begin to see that if we want to reduce the level of violence in our society and help prevent the risk of future violence in those who either impulsively or intentionally committed previous acts of violence, we need a powerful new tool. I hope they will explore the possibility that just as groups and societies can breed violent aggression, there may be a way to develop more constructive methods of expression. There is a great need for research into early childhood intervention, and for methods that are effective in rehabilitating offenders, methods that will certainly need to be individually tailored depending on the individual's biological and genetic makeup and major negative environmental influences. Such methods will certainly be criticized as expensive and time-consuming, which they will be. So is death and incarceration. The question then becomes, how long are we willing to stand current conditions? Genetic determinism is unfounded when it comes to complex behavior. And this is what law students need to know: Only a sophisticated understanding of the interplay of biology, environment, and genetics will help solve these crucial legal issues.

Notes

1. See *Barefoot v. Estelle*, 463 U.S. 880 (1983) (citing amicus brief of the American Psychiatric Association).

2. At least since the late nineteenth century, courts and prisons have attempted to discriminate between the innately criminal, and those who acted merely by force of circumstance (whose crimes, being caused by circumstance rather than nature, would not pose a future danger to society). See, e.g., Stephen J. Gould, *The Mismeasure of Man*, 153–72 (1996) (discussing the influence of ideas about heredity on criminal law).

3. See, e.g., Debra Niehoff, *The Biology of Violence: How Understanding the Brain, Behavior, and Environment Can Break the Vicious Circle of Aggression*, 238 (1999) (noting that Scandinavian twin and adoption studies are often cited favor a role for genetic influence on crime).

4. See Avshalom Caspi et al., Role of Genotype in the Cycle of Violence in Maltreated Children, 297 *Science* 851–53 (2002) (studying 442 men in New Zealand for differences in MAOA activity alleles and correlating these differences with maltreatment in childhood and subsequent violent behavior). The results demonstrated that the high-activity form of the gene did not manifest in violent propensities even if the men had been mistreated as boys, while those with the low-activity form of the gene, who had been mistreated, committed four times as many rapes, assaults, and robberies as the average. A second study by the Caspi group in 2003 also reported on a gene–environment interaction, this time in the promoter region of the serotonin transporter gene: Avshalom Caspi et al., Influence of Life Stress on Depression: Moderation by a Polymorphism in the 5-HTT Gene, 301 *Science* 386–89 (2003) (reporting on the interaction between the short allele of the serotonin transporter gene and stressful environment).

5. Clarence Darrow, The Crime of Compulsion, Address of Case Summation Before John R. Caverly, Chief Justice of the Criminal Court of Cook County (Aug. 22, 1924) in *Attorney for the Damned: Clarence Darrow in the Courtroom*, 65–66 (Arthur Weinberg, ed., University of Chicago Press 1989) (1957).

6. See, e.g., *State v. Davis*, No. M1999-02496-CCA-R3-CD, 2001 Tenn. Crim. App. LEXIS 341 (Tenn. Crim. App. May 8, 2001) (claims of genetic predisposition to mental illness offered to rebut mens rea unsuccessful).

7. See *People v. Tanner*, 91 Cal. Rptr. 656 (Cal. Ct. App. 1970) (neither the link to aggressive behavior nor a chromosomal contribution to legal insanity were established); *Millard v. State*, 261 A.2d 227 (Md. Ct. App. 1970) (upholding trial court's refusal to submit the issue to the jury because the expert failed to demonstrate a link between the XYY condition and the legal definition of insanity); *People*

v. Yukl, 372 N.Y.S. 2d 313 (N.Y. Sup. Ct. 1975) (refusing to order genetic testing or to permit defendant's father to pay for genetic testing because the evidence of a genetic link to violence was not reliably established); *State v. Roberts*, 544 P.2d 754 (Wash. App. 1976) (affirming trial court's denial of genetic testing because of the uncertain causal connection between XYY and criminal conduct).

8. See, e.g., *United States v. Moore*, 486 F.2d 1139 (D.C. Cir. 1973) (finding that a genetic predisposition to addiction did not render conduct involuntary).

9. A possible exception to the failure of these claims is *Crook v. State*, 813 So. 2d 68 (Fla. 2002) (vacating death sentence for failing to consider the defendant's organic brain disease as a mitigating factor), 908 So. 2d 350 (Fla. 2005) (vacating death sentence after resentencing in light of mitigating factors).

10. See *Turpin v. Mobley*, 502 S.E.2d 458 (Ga. 1998) (finding no ineffective assistance of counsel in failing to accept defendant's father's offer to pay for genetic testing for MAOA deficiency analysis after the trial court refused to pay for it).

11. *Landrigan v. Schriro*, 441 F.3d 638 (9th Cir. 2006) (*en banc* rehearing of 397 F.3d 1235), cert granted, Schriro v. Landrigan, S. Ct., 2006 WL 1591780 (U.S. Sept. 26, 2006).

12. *Landrigan v. Stewart*, 272 F.3d 1221, 1224 (9th Cir. 2001).

13. *Idem* at 1228.

14. *Landrigan*, 441 F.3d at 645.

15. *Idem* at 649.

16. Although Landrigan was sentenced by a judge, in *Ring v. Arizona*, 536 U.S. 584 (2002), the Supreme Court ruled that the capital sentencing decision must be made by a jury.

17. See John H. Blume et al., Future Dangerousness in Capital Cases: Always "At Issue," 86 *Cornell L. Rev.* 397, 398 (2001) (observing, on the basis of interviews with over a hundred capital jurors, that "future dangerousness is in the minds of most capital jurors, and is thus "at issue" in virtually all capital trials, no matter what the prosecution says or does not say").

18. See, e.g., Stephen P. Garvey, Aggravation and Mitigation in Capital Cases: What Do Jurors Think?, 98 *Colum. L. Rev.* 1538, 1559 (1998) (citing studies emphasizing the "pervasive role future dangerousness testimony plays in and on the minds of capital sentencing jurors").

19. See Blume et al., *supra* note 123, at 404 (reporting that "even in cases in which the prosecution's evidence and argument at the penalty phase did 'not at all' emphasize the defendant's future dangerousness, jurors who believed the defendant would be released in under twenty years if not sentenced to death were still more likely to cast their final vote for death than were jurors who thought the alternative to death was twenty years or more"). Indeed, it was the explicit recognition of the importance the jury gives to future dangerousness that motivated the Supreme Court to rule that defendants have a constitutional right to be informed of a death penalty alternative if the prosecution alleged future danger as an aggravating circumstance and the alternative is life in prison without parole. *Simmons v. South Carolina*, 512 U.S. 154 (1994); see also *Kelley v. South Carolina*, 534 U.S. 246 (2002) (reiterating the Court's earlier holding in Simmons); cf. *Garvey*, *supra*

note 124, at 1559 (observing that future dangerousness "appears to be one of the primary determinants of capital-sentencing juries").

20. *Gregg v. Georgia*, 428 U.S. 153, 195 (1976).

21. Expert testimony is frequently proffered at capital sentencing proceedings. For example, in the Capital Jury Project, funded by the National Science Foundation, the California portion of the study examined 36 death penalty cases, and found that the prosecution called an expert in 81 percent of the cases, and the defense called an expert in ninety percent. Scott E. Sundby, The Jury as Critic: An Empirical Look at How Capital Juries Perceive Expert and Lay Testimony, 83 *Va. L. Rev.* 1109, 1119 (1997) (noting that "conventional practice at the penalty phase involves presenting an expert to the jury at some point—sometimes more than one—who will testify based upon an expertise gained through training and study").

22. See Vernon L. Quinsey et al., *Violent Offenders: Appraising and Managing Risk*, 62 (1998) (noting that "laypersons and the clinicians had few differences of opinion" about assessments of dangerousness, and that neither had much accuracy).

23. See generally Christopher Webster et al., *The Violence Prediction Scheme: Assessing Dangerousness in High Risk Men* (1994) (detailing studies demonstrating the inaccuracy of violence risk predictions); John Monahan, *Predicting Violent Behavior: An Assessment of Clinical Techniques*, 1 (1981) (surveying the major studies of clinical prediction of future dangerousness and finding the results to show that psychiatrists had about a one in three chance of predicting future dangerousness correctly).

24. See, e.g., Charles W. Lidz et al., The Accuracy of Predictions of Violence to Others, 269 *J. Am. Med. Ass'n* 1007 (1993) (concluding that "clinicians are relatively inaccurate predictors of violence"). In this study, when clinicians divided institutionalized men into two groups, "violent" and "nonviolent," and examined their behavior more than three years later, 53 percent of the "violent" group had committed acts of violence, as opposed to 36 percent of the "nonviolent" group. *Idem*. Random predictions would have a sensitivity and specificity of 50 percent. *Idem* at 1009. Thus, while the results are better than chance, the low sensitivity and specificity of the predictions show "substantial room for improvement." *Idem*. Sensitivity is the percentage of times that a test correctly gives a positive result when the individual tested actually has the characteristic in question. Bruce R. Parker & Anthony F. Vittoria, Debunking Junk Science: Techniques for Effective Use of Biostatistics, 66 *Def. Couns. J.* 33, 34 (1999). Specificity is the percentage of times a test correctly reports that a person does not have the characteristic under investigation. Parker & Vittoria at 34. Actuarial studies, though more accurate than clinical predictions, still predict with less than stellar accuracy: When scores on the most accurate of the actuarial instruments, the VRAG, "were dichotomized into "high" and "low" risk groups, the results indicated that at most, 55% of the 'high scoring' subjects committed violent recidivism, compared with 19% of the 'low scoring' group." John Monahan, Violence Risk Assessment: Scientific Validity and Evidentiary Admissibility, 57 *Wash. & Lee L. Rev.* 901, 907 (2000); see also *infra* Part IV.B.2.c (discussing the VRAG).

25. For a more complete discussion of how the biology of violence and sexual violence ought to inform dangerousness predictions, see generally Erica Beecher-Monas & Edgar Garcia-Rill, Genetic Predictions of Future Dangerousness: Is There a Blueprint for Violence?, 69 *L. & Contemp. Problems* 304 (2006).

26. For this section of the seminar, I typically assign reading from Martin Daly & Margo Wilson, *Homicide* (1988), as well as other selected readings on evolutionary biology.

27. A good introduction to this material is found in Niehoff, *supra* note 4 at 174–81.

28. John Tooby & Leda Cosmides, On the Universality of Human Nature and the Uniqueness of the Individual: The Role of Genetics and Adaptation, 58 *J. Personality* 17, 53–58 (1990).

29. See Avshalom Caspi et al., Role of Genotype in the Cycle of Violence in Maltreated Children, 297 *Science* 851,53 (2002) (studying 442 men in New Zealand for differences in MAOA activity alleles and correlating these differences with maltreatment in childhood and subsequent violent behavior). The results demonstrated that the high activity form of the gene did not manifest in violent propensities even if the men had been mistreated as boys, while those with the low-active form of the gene, who had been mistreated, committed four times as many rapes, assaults, and robberies as the average. *Idem.*

30. Frederick S. vom Saal, Models of Early Hormonal Effects on Intrasex Aggression in Mice, in *Hormones and Aggressive Behavior*, 197, 198 (Bruce B. Svare, ed., 1983).

31. Adrian Raine et al., Brain Abnormalities in Murderers Indicated by Positron Emission Tomography, 42 *Biological Psychiatry* 495 (1997); Adrian Raine et al., Selective Reductions in Prefrontal Glucose Metabolism in Murders, 36 *Biological Psychiatry* 365 (1994).

32. Excellent, accessible reading (even for nonscientists), along these lines, can be found in Niehoff, *supra*.

33. See Beecher-Monas & Garcia-Rill, *supra* note 9, at 334.

34. An example of these views can be found in Richard Dawkins, *The Selfish Gene* (1976).

35. Matt Ridley, *The Agile Gene: How Nature Turns on Nurture* (2003).

36. Richard Lewontin, *The Triple Helix: Gene, Organism and Environment* (1998).

37. Debra Niehoff, *The Biology of Violence: How Understanding the Brain, Behavior and Environment Can Break the Vicious Circle of Aggression* (1998).

38. See, e.g., Beecher-Monas & Garcia-Rill, *supra* note 9 at 337–339.

Designer Genes

Teaching the Ethics of Genetic Research with Clustered Courses

Bethany Hicok and Joshua Corrette-Bennett
WESTMINSTER COLLEGE

Part of the fascination of the new genetics concerns the questions it raises about the construction of knowledge—how, for whom, and for what is this knowledge being constructed?

—ALICE WEXLER, *Mapping Fate*

We have within our grasp the future of mankind, and as things are going the future will be chosen according to the same criteria as people now choose silicone breast implants and liposuction and hair transplants. It will be eugenics by consumer choice, the eugenics of the marketplace. All masquerading as freedom.

—SIMON MAWER, *Mendel's Dwarf*

In his book *High Noon: 20 Global Problems, 20 Years to Solve Them*, J. F. Rischard notes that the biotechnology issues raised by the Human Genome Project are some of the most pressing global issues we face today (Rischard, 2003). After researchers sequenced over 95 percent of the human genome, Francis Collins and colleagues wrote, "In conclusion, the successful completion this month of all of the original goals of the HGP emboldens the launch of a new phase for genomics research, to explore the remarkable landscape of opportunity that now opens up before us. Like Shakespeare, we are inclined to say, 'what's past is prologue'" (Collins, Green, Guttmacher, & Guyer, 2003). Questions about how we define race, gender, disease, and disability become even more pressing when it becomes possible for us to select what traits society deems more "desirable." While most would agree that the ethical, legal, and social issues raised by the Human Genome Project might be best addressed by drawing on expertise from multiple disciplines, the challenge for higher education in teaching this expansive topic is how to draw on sufficient expertise; most faculty members, after all, are narrowly educated in a specific field. Courses where a single professor takes an interdisciplinary approach, therefore, can be quite limited in their scope or depth.

Westminster College has addressed a number of these weaknesses with an innovative approach that enhances the liberal arts experience for students as well as faculty: the concept of clustered courses. Unlike other institutions where the cluster can consist of a grab-bag of courses that form a minor or concentration, Westminster's cluster consists of two interlinked, integrated, four-credit courses that students are required to sign up for in the same semester. This approach to interdisciplinary learning, combined with what we gained through the Dartmouth Ethics Institute in the summer of 2004, allowed us to develop our cluster course, "Designer Genes," which we taught in the spring of 2005 and will offer again in 2007. Our cluster consists of an introductory laboratory course in biology, "The Science and Ethics of Human Genome Research," and one in literature, "Genetics in Literature, Film, and Culture." Students who took the cluster in 2005 came from a broad range of disciplines, including the sciences, sociology, music, English, international business, and elementary education.

The title of our cluster, of course, plays on multiple levels of meaning and already begins to hint at the central issues that serve as the foundation of this cluster. Because literary and film analysis focus primarily on language and representation, it is a discipline that is well suited to getting at the social, ethical, and scientific complexities of this issue. After sequencing the genetic code of an individual, we are still left with ethical dilemmas and enduring questions. What do we know? How do we know it? Will this information be used to enhance the individual (or society)? If so, how will it improve the individual (or society)? Who should make those decisions? Will we be able to design our own genes, creating designer babies and societies? Advertising has already picked up on the possibilities of such genetic determinism. A recent *New York Times* magazine piece on film director Sophia Coppola announces: "In this brave new world where science has enabled short people to become tall and homely people to become 'pretty,' isolating the style gene is only a matter of time" (Hirschberg, September 19, 2004). In another recent article, Jonathan Rockoff points out there are already hundreds of genetic tests on the market for diagnosing disease, and advertising seems to be driving the demand (Rockoff, November 30, 2005). The cluster allows us to explore these and other issues associated with the Human Genome Project in considerable depth both within and across disciplines. In this chapter we address the pedagogical value of the cluster and the various teaching methods, lab experiments, and assignments used to generate discussion, debate the compelling questions raised by the new genetics, and foster critical thinking on the subject.

Overview of the Cluster

Rather than take a broad survey approach, we organized the integrated courses around topics—family, risk and inheritance; race and genetic diversity; genes and mutations; eugenics; forensics; sexuality; and cloning technologies—that would provide us with significant overlap between literature and biology. For each topic, Corrette-Bennett taught the basic biological principles and ethical issues surrounding human genome research, while Hicok explored the same topic through literary and film texts, such as Alice Wexler's memoir *Mapping Fate*, Simon Mawer's novel *Mendel's Dwarf*, and Andrew Niccol's 1997 film *Gattaca*. We organized our lecture and laboratory schedules so that we could attend each other's classes and/or labs whenever possible. There were several advantages to participating in each other's classes. First, because the two courses approached the subject matter from distinctly different perspectives, we each experienced the other faculty member's course in a manner similar to that of a student: an English professor learning about restriction enzymes and analyzing DNA from her own cheek cells; a biology professor learning about epigraphs, while analyzing the social and emotional complications of inherited disorders. Second, our occasional participation in class discussion provided alternative and sometimes conflicting perspectives and allowed us to raise issues that the students might have been too timid to broach.

In both classes, we supplemented the primary material with essays from philosophy, economics, and law (drawn primarily from ELSI materials) in order to point out that the ethical dimensions of human genome research are not just about science and literature. We used a variety of pedagogical methods to help students explore the issues, including short written reflections, weekly laboratory exercises that complemented texts and lectures, case studies, debates, a mock trial, board games designed by the students, guest speakers, and a final group project that required the students to integrate concepts and ideas from both science and literature.

Family, Risk, and Inheritance

Because interconnected themes were such an important feature of our cluster, we began with one of the most enduring and powerful: "family, risk and inheritance." Alice Wexler's memoir about her own family's struggle with Huntington's disease served as a superb sounding board for both classes because the memoir itself is interdisciplinary. Wexler is a historian writing about science, and her book exemplifies Rischard's point about the importance of experts coming together from various fields to find solutions. Wexler describes a family racing against time to reveal possible genetic secrets in hopes of saving their mother, only to realize that they are also trying to save themselves and their family. While Alice Wexler's therapy takes the form of literature and history, her sister Nancy looks for answers using the scientific approach, eventually becoming a prominent figure in Huntington's research. Wexler points out that race, class, and gender all play a role in who has the power to speak when it comes to these issues, and Wexler's account helps students see that knowledge—indeed, the field of "biology itself"—is "shaped by its social, political, and cultural contexts" (Wexler, 1996).

Wexler's is such a rich, multilayered text that it is also a perfect one for introducing students to close reading strategies, which require them not to just read for information but to analyze passages of text to understand their deeper meaning and significance. *Mapping Fate* sets up many of the major questions and issues of the cluster. Wexler's book humanizes the social stigma of inherited disorders and helps us to think about important questions related to human genome research: What is the basis of inherited disorders such as Huntington's and why is it heritable? Why do people choose a career in science? How do we use scientific knowledge once we acquire it? Do genetic diseases define who we are? Such questions stimulate student discussion and interest, which subsequently carries over into the biology course and motivates them to learn more about the basics of Mendelian genetics, DNA, inherited disorders, and probability. Because Wexler's book also gets into the complicated issues of familial inherited disorders and disease, we had students isolate DNA from their own cheek cells and then analyze them for the presence or absence of genetic markers using polymerase chain reaction (PCR) amplification. We asked students to write short two-page reflection papers on the books we read, enabling them to explore more freely some of the often emotional and controversial territory this topic elicits. For one student, who had been struggling with mental illness, Wexler's book spoke directly to her own experience of trying to convince her parents that her identity was more than just her mental illness.

Race and Genetic Diversity

The issues of family and culture raised by Wexler led nicely into our next topic: race and genetic diversity. The text selected for this section does not really delve into the genetic aspects of race, but nevertheless provides a compelling introduction to the social perceptions and construction of race, beauty, and identity. Toni Morrison's 1970 novel *The Bluest Eye* is a chilling, modern retelling of Mary Shelley's *Frankenstein*. The novel is about growing up African American in the 1940s in a pre–Civil Rights America where the white, blond-haired Shirley Temple is the icon for little girls. One of the novel's major themes is racial self-loathing and how prejudice and institutional discrimination contribute to the destruction of the book's main character, Pecola Breedlove. Pecola wants only one thing—blue eyes—because she thinks that will make her beautiful. Again, the issues of identity and choice are prominent here. Morrison uses excerpts from the *Dick and Jane* readers of the 1930s and 1940s, which featured only White characters, to frame the story and illustrate how the main characters are trapped by cultural representations in which they cannot see themselves (Morrison, 2000).

While the students were grappling with the social issues of diversity and race in the literature course, they were also introduced to the scientific factors that define genetic diversity and physical appearance in the biology course. Students learned about genes, proteins, and the cellular pathways that determine physical features, such as skin color. They were also presented with a geographic distribution of skin color from around the world and asked to consider environmental factors that might contribute to this distribution (Iqbal, November 2002). The purpose of this approach was to have students compare and contrast scientific knowledge and perceptions of phenotypic differences (and similarities) between human populations with social understandings and perceptions of race. When the material was presented in this manner, students quickly realized that current racial classifications are simply the product of outdated social constructs, and that physical features currently used to distinguish between two races (such as skin color) are the result of minute genetic differences, which pale in comparison to the amount of genetic diversity present within a single race or ethnicity.

Genes and Mutations

Simon Mawer's novel *Mendel's Dwarf* provided our cluster with an important transition from basic variations within human populations to more dramatic manifestations of variations, such as inherited disorders and diseases. Variations such as skin color are all too often mistakenly seen by the general population in static terms—that is, physical traits that never change and are simply passed on from one generation to the next. Human disorders and diseases provide a much more dynamic example, both in cause and effect, and are inherited by some generations, while spontaneously occurring in others. Disorders and disease pique the interest of students for a variety of reasons and give rise to a host of ethical and social dilemmas. The main character of Simon Mawer's excellent novel is Benedict Lambert, a geneticist who is also a dwarf, and, as he says early in the novel, "I search for the gene that caused me" (Mawer, 1998). Like many other things in our culture, Benedict Lambert claims, the choices opened up by the new genetics will be ruled by the marketplace—that is, "eugenics by consumer choice." Like Wexler's memoir, Mawer's novel is not just *about* genetics. The novel's intertwined narratives—the story of Benedict Lambert and the history of Gregor Mendel's genetic discoveries—form a kind of literary double helix as the story unfolds.

Mawer's novel provides sufficient information for a detailed analysis of the genetic cause and physical effects of inherited disorders, such as achondroplasia (dwarfism). Mawer, who himself was a biology teacher, also uses genetic language to organize the novel's structure, so that chapter headings like "Restriction" have multiple meanings that were investigated in both courses of the cluster. Because our students have already been introduced to restriction enzymes in the biology course, when Benedict Lambert gives a talk on eugenics, the students know that just as a restriction enzyme cuts and selectively removes genetic information, inherited disorders and eugenics can selectively cut out, or remove, individuals from a population.

The novel also serves as a springboard for fundamental biological topics, such as spontaneous mutation, natural selection, and artificial (or social) selection, providing once again an opportunity for interdisciplinary inquiry. Because the text for this part of our cluster was dense with meaning, we divided students into small groups during the literature class and focused on close reading techniques. We made each group responsible for several chapters of the novel, which they then had to discuss and present to their peers. Laboratory exercises for this section of the cluster included researching a selected genetic disorder using the On-Line Mendelian Inheritance in Man (OMIM) database (McKusick), electrophoretic separation of hemoglobin proteins from normal individuals and those suffering from sickle cell anemia, and mutagenesis of bacteria with ultraviolet light.

Eugenics

One of the cluster's major themes is the genetics of identity: Are we defined by our genes? How much control do we have over our physical and intellectual development? Is information from the Human Genome Project leading to more useful discoveries or a new type of social discrimination? Has the line between natural and artificial selection been erased? We used the film *Gattaca* in order to emphasize the concepts of artificial selection, eugenics, and genetically engineered societies. As most ELSI participants already know, the subject matter of this film is useful for discussing the advantages and disadvantages of a society where one is primarily defined by one's genes, a place where, as the main character, Vincent, says, "We now have discrimination down to a science" (Niccol, 1997). As with other texts used for this cluster, *Gattaca* features genetics not just as a subject, but also as a structuring element that adds layers of meaning to the film. Genetic metaphors abound—the staircase in Jerome's house takes the form of a double helix; the film's title and title sequence is formulated from repetitions of the four DNA bases, GATC—guanosine, adenosine, thymine, and cytosine; the terms "borrowed ladder" and "degenerate" (pronounced de-gene-erat) identify individuals, such as Vincent, who use other people's genetic identity to advance in society or who are unworthy of society's notice; and so on.

The film forcefully drives home the point that in this society, one's DNA defines one's identity, and so when Vincent takes on the identity of the genetically enhanced Jerome in order to get into the space agency known as Gattaca, we see him literally trying to scrub off and wash away his own "in-valid" self every morning in order to adopt the new identity. In one powerful scene, Vincent is on the beach, hunched almost into a fetal position, using the sand and salt water to scrub off his own dead skin cells. The scene, by reenacting a ritualistic baptism of washing away the sinful self in order that the individual can be cleansed and reborn, makes clear the high price that individuals must pay in this society in order to be "acceptable." As David Kirby has argued, *Gattaca* is a full-scale "attack on genetic determinism," and the filmmakers themselves "act as bioethicists," challenging us to confront "the consequences of unrestricted human-gene therapy" (Kirby, 2000). The film comes down heavily on the side of natural selection in its use of symbolism. To take just one example, the sea, the natural origin of life, becomes the scene of combat between the two brothers, Vincent and Anton, one a "love child," the other, genetically selected; on a symbolic level these competitions serve to reinforce the idea that Vincent and the film itself are literally fighting for the survival of "natural selection."

We introduced students to some basic elements of formal film analysis to help them to understand that learning to "read" visual texts can enhance a film's overall message and affect on the viewer. The use of high-angle shots in *Gattaca*, for example, reinforces the sense of social control the film is trying to convey. In order for the students to evaluate the scientific concepts presented in the film, we introduced them to a number of reproductive technologies currently available. One particularly memorable scene in the movie involves the conception of Vincent's brother Anton. Unlike Vincent's conception and birth using natural means, Vincent's parents decide that his younger brother should be conceived using in vitro fertilization followed by preimplantation genetic diagnosis (PGD), or genetic screening. The scene with his parents staring at the TV monitor and selecting the optimal embryo makes the process appear cold and sterile, foreshadowing things to come. But we point out to the students that these techniques have already been practiced for a number of years, significantly reducing the incidence of miscarriage and lethal childhood disorders, such as Tay–Sachs. The film also introduces a number of forensic sampling methods and genetic tests, which led into our section on forensics.

Sexuality and Genetics

The topic of sexuality and gender is one where classroom teaching can have a profound effect on students' preconceived assumptions, particularly when combining an examination of the social issues with a review of basic human anatomy, physiology, and sexuality. In this section of the cluster, we thought it was crucial for students to learn about important stages of human embryogenesis and development, such as the genetic and cellular causes of primary and secondary sexual characteristics. Students were then introduced to a number of well-documented variations in human sexual development, such as Turner syndrome, Klinefelter syndrome, hermaphrodites, and androgen insensitivity. Once students were exposed to the physical complexities of human sexuality, they were more amenable to discussing research and literature that explores sexuality and gender. We began by reading the preface and Chapter One of Dean Hamer and Peter Copeland's book *The Science of Desire: The Search for the Gay Gene and the Biology of Behavior* (Hamer & Copeland, 1994). A popular account of Hamer's scientific research, this book reads like a good detective novel in some instances, and it is useful for furthering our discussion of the ethics and politics of scientific research, which began with Wexler's memoir. We also had students read the original scientific paper "A Linkage Between DNA Markers on the X Chromosome and Male Sexual Orientation," published in the journal *Science*, which is included in the appendix of Hamer's book. We discussed the experience of reading the different accounts, the difference in style, level of discourse, content, and in target audience. This exercise helped reinforce the goals of a similar information literacy assignment used earlier in the semester. These assignments were designed to introduce students to sources of information; allow them to assess the scientific community's ability to educate the public; develop the ability to evaluate sources of information; and finally, to identify the origins and constant evolution of scientific information.

We also asked the students whether they thought scientists working in this field should recuse themselves from the ethical issues raised by their research. Even at the end of their scientific article, Hamer et al. are quick to consider the ethical dimensions of their research, issuing a strong caveat against what they consider to be the misuse of such knowledge: "We believe that it would be fundamentally unethical," they write, "to use such information to try to assess or alter a person's current or future sexual orientation, either heterosexual or homosexual, or other normal attributes of human behavior" (Hamer & Copeland, 1994). The students also read Edward Stein's "Choosing the Sexual Orientation of Children" (Stein, 1998) and C. L. Ten's "The Use of Reproductive Technologies in Selecting the Sexual Orientation, the Race, and the Sex of Children" (Ten, 1998). By carefully considering some of the ethical arguments for and against sexual selection, students began to understand how complicated these choices can be.

We concluded this section with a mock Supreme Court case, the case of Baby XY, which involved a couple, George and Susie Lyon, who found out through genetic testing that Susie was carrying a child with the "gay gene." The Lyons argued that a Pennsylvania law that prohibited abortion on the basis of sex or sexual orientation was unconstitutional because of the precedent set by *Roe v. Wade* that guarantees the reproductive rights of the mother. The students were assigned roles in teams. The team of lawyers for the Lyons argued that Susie Lyon had the right to terminate her pregnancy. Lawyers for the state argued that there were certain instances that should be made exceptions when it comes to *Roe v. Wade* and reproductive rights. The rest of the students acted as the Supreme Court and had to write a majority opinion once the case had been heard. The mock trial allowed students to bring together much of what they had learned about the ethical and legal issues involved and to test that knowledge. Most importantly, it forced students to argue positions in which they didn't necessarily believe, sometimes leading them to change their minds, or at least disrupting their certainties for a few moments.

Cloning Technologies

In the final section of the course, we considered cloning technologies, and students were responsible for reading and preparing various articles on cloning to enable them to lead class discussions. Articles included Michael J. Sandel's anti-cloning article "The Case Against Perfection," which appeared in the *Atlantic Monthly* (Sandel, April 2004), and on the other side, Ron Green's "I, Clone," published in *Scientific American* (Green, fall 1999). We rounded this out with a couple of different views presented by the President's Council on Bioethics in its report on human cloning. We asked students to choose two or three of the main arguments to evaluate and critique. They had to demonstrate where they saw biases or problems in logic and to look closely at the kind of language the writer used to appeal to his or her audience. Again, the point here was to give students more of an opportunity to wrestle with the issues in depth, while minimizing instructor-directed discussion. We used laboratory time to review current and realistic future cloning technologies discussed in lectures and then asked the students to develop a board game that addressed the scientific, social, and ethical issues presented by cloning. At the end of the semester, each group played the other groups' games and then assessed them based on a number of criteria: creativity, scientific accuracy, educational and entertainment value, and how well they addressed the most relevant issues. The exercise proved to be a very useful means of assessing the students' knowledge of this topic.

Final Project

We developed a final project that allowed students to bring the two disciplines together and for us to assess, at least in part, the success of the cluster; how much they had learned about the science; how well they understood the texts; and how well they could address the complexities of the issues. The project, however, had to combine the approaches, knowledge, and methodologies of both disciplines. The students were allowed to select from a number of topics for their final projects. We asked them to work in groups of two or three students and to either research or debate an issue covered within the cluster. If they chose the debate, each member of the group would look at a slightly different aspect of the debate. They had to define the issue, present relevant research from a variety of sources that identified the scope of the issue, and then evaluate and document their supporting resources. If a group chose a text that we had not covered during the cluster, its presentation had to introduce the film or literary text, discuss significant themes, and provide an interpretation of the text based on what the members had learned in both classes about genetics and literary analysis. One particularly interesting project involved in innovative approach for looking at genes, inheritance, royalty, and art through a study of portraits of the Austrian Habsburg family. This project looked at the impact of genetic inbreeding on the physical characteristics and health of a ruling family (the famous Habsburg jaw and other disorders associated with this trait) and its subsequent influence on politics and history. These students not only drew on history and science, but they also used portraits of the royal family that showed the increasing elongation of the jaw as the trait was passed on from generation to generation.

Conclusions

Reflections from students are probably the most encouraging indicators of the success of this cluster. A music major, for instance, who had concerns about the scientific requirements of the Westminster curriculum said, "[The cluster] has show[n] me how one discipline can help me understand another better. The books we have read in English have provided an easier understanding of the science half of the course. These books explained the science in a way that anyone reading the book could at least grasp in a simpler form." One of the biology majors mirrored the previous student's sentiment, but from a scientist's perspective: "Before I experienced this literature course, I only looked at the clinical side of disease," she wrote. But after reading Alice Wexler's memoir, she said she had "come to see the person as well as the sickness."

The two courses are designed to give students access to multiple languages for looking at and comprehending these issues.[1] Students and participating faculty are exposed to various ways of knowing and understanding current topics that directly influence their lives. The following student reflection epitomizes the intent of this cluster experience:

> The biology course not only encouraged me to learn more about the issues that we dealt with in class, but it has encouraged me to stay updated and continue learning more about the Human Genome Project in the future, and the effects that it has on me, as well as those around me. The English section of the course put all of the biology into perspective for me, and provided real life, tangible experiences and examples. I feel this section of the course had the greatest effect on me. It dealt with a lot of social issues and concerns that surround those who are part of the Human Genome Project, as well as those who can and will be affected, as research continues. The course opened my eyes to see and understand . . . the importance of being educated about the Human Genome Project.

Note

1. We understand that those faculty members who explore the science and ethics of human genome research often do not have the luxury of up to nine hours per week of contact time with students. So we recommend the following for humanities faculty who teach this topic with only three hours per week of student contact time: the use of a topic-oriented approach to convey concepts and discuss issues; texts that promote a balance of science, ethics, and literary analysis, such as Alice Wexler's *Mapping Fate*, Simon Mawer's *Mendel's Dwarf*, and the film *Gattaca*; use of the text to discuss scientific concepts as needed; a final culminating project along with pre- and postreflection papers for assessment purposes; a mock trial or guest speaker, if time permits.

References

Collins, F. S., E. D. Green, A. E. Guttmacher, and M. S. Guyer (2003). A Vision for the future of genomics research: A blueprint for the genomic era. *Nature, 422*, 835–847.

Green, R. M. (Fall 1999). I, Clone. *Scientific American, 10*(3), 80–83.

Hamer, D., & Copeland, P. (1994). *The Science of Desire: The Search for the Gay Gene and the Biology of Behavior*. New York: Simon and Schuster.

Hirschberg, L. (September 19, 2004). The Originals. *The New York Times*, pp. 147–149.
Iqbal, S. (November 2002). A New Light on Skin Color. *National Geographic Magazine: Online Extra*, http://magma.nationalgeographic.com/ngm/0211/feature2/online_extra.html.
Kirby, D. A. (2000). The New Eugenics in Cinema: Genetic Determinism and Gene Therapy in Gattaca. *Science Fiction Studies*, *27*, 193–215.
Mawer, S. (1998). *Mendel's Dwarf*. New York: Penguin.
McKusick, V. A., et al. OMIMTM—Online Mendelian Inheritance in Man™. Johns Hopkins Medicine. The McKusick–Nathans Institute of Genetic Medicine. Retrieved in 2005 from the National Center for Biotechnology Information web site: http://www.ncbi.nlm.nih.gov/entrez/query.fcgi?db=OMIM.
Morrison, T. (2000). *The Bluest Eye*. New York: Plume.
Niccol, A. (Writer and Director). (1997). *Gattaca*. Sony Pictures.
Rischard, J. F. (2003). *High Noon: 20 Global Problems, 20 Years to Solve Them*. New York: Basic Books.
Rockoff, J. D. (November 30, 2005). Surge in genetic testing raises quality concerns. *Baltimore Sun*, 2A.
Sandel, M. J. (April 2004). The Case Against Perfection. *Atlantic Monthly*, 51–62.
Stein, E. (1998). Choosing the Sexual Orientation of Children. *Bioethics*, *12*(1), 1–24.
Ten, C. L. (1998). The Use of Reproductive Technologies in Selecting the Sexual Orientation, the Race, and the Sex of Children. *Bioethics*, *12*(1), 45–48.
Wexler, A. (1996). *Mapping Fate: A Memoir of Family, Risk, and Genetic Research*. Berkeley: University of California Press.

Nature and Culture

Teaching ELSI in a History-of-Science Course

Myles W. Jackson
POLYTECHNIC UNIVERSITY

Historians of science have a particularly daunting (and infinitely fascinating) task: We investigate how science shapes society and culture, while simultaneously piecing together how society and culture frame science. Demarcating precisely what constituted science and what counted as culture is difficult, as the boundaries between these two spheres of human activity are often blurred. Science is generally assumed to be the antithesis of culture: objective, disinterested, and unchanging (and hence, technically, incapable of possessing a history). Culture, on the other hand, is deemed to be subjective, relative, and interested. Science deals with nature, while culture is quintessentially about people. This course challenges that simplistic, hackneyed, and historically inaccurate depiction of these two domains.

There are a number of tools at the historian's disposal to analyze the relationship among biology, society, and culture. Looking at numerous case studies, my course draws upon economic, social, intellectual, and cultural histories in order to probe the contours of biology and society from the French Enlightenment to the present. Students need to appreciate that the way in which we talk about the past shades the way in which we understand and depict our present and helps frame the types of discussions we shall have in the future. Contrary to the view of being relegated to bow-tie-wearing octogenarians who provide quaint cocktail-party stories, historians actually act as the gatekeepers of our collective memory.

It turns out that it is rather difficult to reconstruct the past. Archives are incomplete, often representing the predilections of previous historians, collectors, or archivists. And, of course, many have deteriorated (or have been destroyed) over time. Many actors have been silenced throughout history, due to race, creed, gender, and/or social class. For example, we know much more about aristocratic men at Italian Renaissance courts than we know about female courtesans. Scientific instrument makers, who were working-class artisans, rarely wrote down their daily routines and practices, as their skills were tightly guarded trade secrets. And one can hardly view a slave owner's account of his slaves as being "disinterested and objective." So historians need to labor a great deal to reconstruct the past, realizing that by living in a different period, they often ask different types of questions than their historical actors might have. They also draw upon different resources to answer those questions. History is an endless dialectical process of living in the present (and trying to explain the present) with recourse to a greatly fragmented past.

Similarly, biologists must work at understanding nature, which has no voice. Indeed, it must be given one (or, as it turns out, a number of them) by scientists, probed by experiments, or represented by models, hypotheses, or theories. And as historians of science, we need to understand historically how scientists (or their predecessors, the experimental natural philosophers) came to understand nature in order to understand how science works today. If history teaches us anything (other than those cocktail-party stories), it is that where we find ourselves today (whether it be social or scientific) is not inevitable, or "natural." Rather, it is a result of the inherently social processes of investigations, negotiations, and conclusions.

Biology, as the science that depicts and explains nature, plays an immensely powerful role in society. This is particularly true of molecular biology over the past thirty years. We are undoubtedly in the age of the gene (Fox Keller, 2000). And molecular biology (particularly immunology) has provided us with nuanced definitions of self. The nature-versus-nurture debate seems to have swung over to the side of nature. Obesity, alcoholism, intelligence, sexual preferences, zest for life, and thrill seeking now all seem to have a genetic component. The ethical, social, and legal implications are enormous. To that end, my course illustrates how the nature-versus-nurture debate is certainly not new, but has enjoyed an impressive (and at times depressive) history. The goal of my seminar is to trace that history over the past three centuries and show how the debate has evolved to the present day. Biology students need to appreciate how their discipline shapes everyday life. Also, they need to realize how the types of questions they ask as well as the ways they go about answering them are social processes. And humanities majors not only need to learn how the social, ethical, and legal implications of recent molecular biological discoveries can affect everyone, but should begin to comprehend how it is that a scientist thinks. Much is at stake: Scientists should not allow the public debate on ELSI to be carried out by antiscience fanatics, and, conversely, it would be unwise for humanities students to surrender ELSI discussions to the scientists themselves.

This course aims at having students debate the issues. They are divided up into small groups comprising three to five students, each containing at least one biology major and one history major. They draw upon each other's expertise. For example, the biology major can explain the science of stem cells or genetic sequencing, while the historian can help the biology student understand the role of the bourgeoisie in defining gender roles in post-Revolutionary France. As a result, students not only learn from me (as I still learn from them), but they teach each other. Since the course only meets for three hours per week, the goal is to infect students with a passion for the field so that they are anxious to discuss the material covered in the lectures and reading outside the classroom. The course also has a blog and e-mail, so that students can engage with the professor and each other outside the classroom.

The course commences with the importance of natural history (which today is better known by its three descendants, zoology, botany, and geology) to social and cultural distinctions in eighteenth-century France (see N. Jardine et al., 1996, and L. J. Jordanova, 1989). Nature was the supreme arbiter of culture. Studies on the reproductive organs of plants and animals (including humans) and on the human musculature and nervous systems were used to naturalize the distinction between the social roles of men and women. Savants, including J. J. Rousseau, who was an avid plant collector and classifier, argued that women were passive and incapable of rational thought as a result of their biology. Hence, their purpose in life was to bear and rear children. Consequently, women were given the responsibility to teach their children the values and morals of the French Republic. Men, on the other hand, were active and rational due to their biological characteristics, and it was argued that they were more suited for the political and economic affairs of the State. These distinctions were certainly not new to the eighteenth century. But now those differences were no longer merely based on the teachings of the Church, but were now bolstered by the status of nature, which had temporarily trumped the Roman Catholic Church as the generator of moral norms. Nature had totally subsumed nurture. Male savants constructed nature to legitimize the preexisting social arrangements. They then proceeded to efface their actions, arguing that nature was simply revealing itself. Objectivity was crucial, particularly since the State argued that they based society on natural law.

The next unit focuses on pre-Darwinian theories of evolution in France and Great Britain. Drawing upon the pioneering work of A. Desmond (1989), students learn that theories of evolution were intricately and inextricably interwoven with theories of religion and the State. Whether or not a species could change over time fundamentally challenged the theory of the Great Chain of Being as well as posing ticklish questions about the role, purpose, and legitimacy of the monarchy and representative democracy. The key to this unit is to stress that the solution to the question about the origins of human life was (is, and probably always will be) simultaneously a biological, theological, and political solution. It would be ahistorical to ask: So what was "science" and what was "mere politics"?

We then discuss Charles Darwin's theory of natural selection as well as the myriad of political and religious responses to his *On the Origin of Species* (1859) and *The Descent of Man* (1871). And we trace how his theories migrated from the natural world into the socioeconomic and political realms. By the late nineteenth century, science was threatening religion as the system governing ethical decisions. Social Darwinism, a phrase coined by the conservative German Darwinian Ernst Haeckel, was an ideology touted by a plethora of political ideologies. It was commonly (and perhaps most accurately) used to support laissez-faire economics. And paradoxically, it formed the cornerstone of a number of theories of anarchism (as promulgated by the Russian Pyotor Kropotkin) as well as extreme forms of racism, as would occur during the Third Reich in Germany. The point I drive home is that there is no inevitable link between social Darwinism and fascism. Scientific theories have been (and still are) used and abused to fit many differing political ideologies, which are often mutually exclusive. Nature is a powerful source of legitimacy. It connotes objectivity, disinterestedness, and "the Truth." Sadly, when science is implemented to marginalize and even silence the powerless, the crime is particularly pernicious. Students begin to see how science began to evolve into a privileged body of knowledge. They often debate the ramifications of such a trend. Topics include the evolutionary epistemology as a model of behavior and historical explanation.

The second half of the course begins with the eugenics movement in Britain. The link between evolutionary theory and eugenics is not a very difficult one to forge. The statistician Francis Galton, the so-called "Father of Eugenics," was Charles Darwin's cousin. And Galton credited his cousin's work with providing the impetus for his theories, which were subsequently refined and championed by the statistician Karl Pearson. Hardly restricted to radical right-wingers, eugenics became a potent weapon in the arsenal of Progressive-Era policymakers and politicians in the United States during the first three decades of the twentieth century. D. J. Kevles (1985) and D. B. Paul (1995 and 1998) provide highly enlightening and readable accounts of the horrors committed in the name of eugenics. For Progressive Liberals, science and statistics (and not religion) clearly indicated the drawbacks of permitting the "less desirables," that is, "profligate women," "the feeble-minded," paupers, "retards," "morons," epileptics, and alcoholics, to procreate. A host of social problems had been collapsed into genetic tendencies, which—it was thought—would be eradicated by forced sterilizations. And the unit concludes with the atrocities committed by the Third Reich, whose officials (particularly the physician and SS officer Josef Mengele) turned to social Darwinism and American eugenics programs to legitimize their heinous practices in concentration camps.

In a somewhat provocative manner, we turn our attention to the Human Genome Project (HGP). We ask, "Does the HGP offer a kinder, gentler form of eugenics?" D. J. Kevles and L. Hood (1992), K. Davies (2001), and J. Shreeve (2004) proffer three excellent accounts of the history of the HGP. D. J. Kevles and L. Hood (1992) also offer compelling essays of ELSI of the HGP. Students read how a number of scientists, including James Watson, wished to allay public fears of a new form of eugenics before the HGP commenced in 1990. After studying the origins of the HGP, we turn to ELSI issues. We discuss the power of genetic tests, which have been developed as a result of the HGP. These genetic tests seem to be specific and precise and therefore objective. Hence, they are very powerful. Genetic essentialism, or reducing social conditions to one's genetic makeup, is an all-too-common phenomenon. It presumes, often fallaciously, that if one has a particular gene or gene constellation, then one will inevitably become afflicted with the disease. But as molecular biologists well know, many diseases, such as hypertension and asthma, result from a combination of a genetic predisposition and environmental factors. Genetic tests also raise another serious issue: Who should be permitted access to the results of those tests? Should employers or insurance companies be informed of your genetic information? In 1995 the State of Oregon passed the first genetic privacy law, which has served as a model for emulation by other states. We study the law and its subsequent revisions, which many viewed as pandering to biotech companies and medical research laboratories in the Portland area. The commercial laboratories accused the initial law of strangling research efforts. Does genetic privacy hinder research? If so, do we need to sacrifice individual privacy? Is there a compromise solution?

We also discuss patenting human genes and gene fragments. This amazing phenomenon also needs to be placed in historical context of scientific inventions and intellectual property. Ever since the early modern period, experimental natural philosophers (and later scientists) have engaged in priority disputes over authorship. Students (including science majors) tend naively to think that science is totally free of commercial interests. Yet there exists a burgeoning literature in the history and sociology of science illustrating how scientists apply for, and receive, patents for their work (see M. Biagioli and P. Galison, 2003). More important, the patenting of scientific inventions and techniques challenges the fundamental notion of the openness of the scientific enterprise. Although in principle patents encourage disclosure of information by guaranteeing protection from potential pirates, numerous scientists have complained that gene patents have actually hampered the free flow of information and collaboration. Students debate what a patent actually means. While some argue that it connotes ownership, most biotech companies vehemently deny the charge that they wish to own a person's genes or gene fragments. They merely wish to protect the vast investments they have committed to research and development. Students relish the chance to debate this topic. They search various web sites in order to glean information on the pros and cons of gene patenting. The key is to have the students themselves debate questions of private versus public funding and research, secrecy, ownership, authorship, intellectual property, and credit.

We continue with these issues when discussing genetically modified organisms (GMOs). As Oregon was the first state to attempt to pass legislation to label foods containing GMOs (the resolution was resoundingly defeated, 25% for and 75% opposed), the topic is of much interest to students in the Northwest. We discuss the topic of agrotech companies such as Monsanto and their role in providing genetically engineered crops to U.S. farmers as well as farmers from a number of developing nations. I have a plant molecular biologist from Willamette offer a guest lecture on the techniques and policies of these companies. And I show the class "Harvest of Fear," an excellent documentary produced in 2001 by *Frontline* and *Nova* on the science and politics of GMOs (http://www.pbs.org/wgbh/harvest/). Once again, lively and often passionate discussions ensue, as some of my students are son and daughters of farmers who use GMOs, while others are militant protectors of the environment.

We conclude the course with the science, ethics, and policy of stem-cell research. We debate the various views of the status of the embryo, the myriad of religious views on the subject, United States and other nations' policies since 1998, and the alternatives to stem cell research. We also explore the various consequences of the very real possibility of leading U.S. scientists leaving the United States to pursue this research in other countries.

In short, this course invites students to look at the historical relationship between nature and culture throughout the past three centuries. In so doing, we do not argue that "history repeats itself." Such a dangerously naïve view assumes that history, *pace* nineteenth-century Prussian scholars, is some organic entity that possesses natural laws. More sinister, such a view effaces the culpability of historical actors. Rather, we see how similar historical debates have framed contemporary discussions concerning ELSI. In a very real sense, ELSI is challenging us to rethink the very essence of science, its character and practice. All of us, scientists and nonscientists alike, need to shed our nineteenth-century notion of what seventeenth-century science is all about. Previous notions of disinterestedness, openness, and authorship are now antiquated. This course is an invitation to all members of the academy to discuss these new issues with a view to make the future just a bit brighter than the present.

Topics for Group Discussions and Papers Throughout the Course

1. How did Enlightenment theories of religion, gender, sex and economy influence Carl von Linné's (Linnaeus') botany?
2. Explain how solutions to questions of theories of evolution were simultaneously solutions to political and religious questions.
3. Explain how social Darwinism used the theory of natural selection to bolster a plethora of evolutionary theories in late-nineteenth-century Europe.
4. You have been voted in as a member of your home state's Board of Education. Do you permit Intelligent Design Theory to be taught? Why or why not?
5. Explain how American politicians and scientists alike appealed to the "scientific, objective, and universal basis" of eugenics to discriminate against people of color and members of the lower social classes.
6. Is the Human Genome Project a kinder, gentler form of eugenics?
7. President George W. Bush's Bioethics Advisory Board on stem-cell research seeks your advice on embryonic stem cell research. What do you say?

8. Do you think that the current level of governmental regulation on genetically modified organisms (GMOs) is sufficient? Why or why not?
9. Does the patenting of human genes, their products, and fragments hinder research?
10. Do genetic privacy issues hinder research in molecular biology?

References

Biagioli, M., and P. Galison. (2003). *Scientific Authorship: Credit and Intellectual Property in Science.* New York and London: Routledge.

Davies, K. (2001). *Cracking the Genome: Inside the Race to Unlock Human DNA.* New York: The Free Press.

Desmond, A. J. (1989). *The Politics of Evolution: Morphology, Medicine, and Reform in Radical London.* Chicago: University of Chicago Press.

Fox Keller, E. (2000). *The Century of the Gene.* Cambridge, MA, and London: Harvard University Press.

Jardine, N., J. A. Secord, and E. C. Spary. (1996). *Cultures of Natural History.* New York and Cambridge, UK: Cambridge University Press.

Jordanova, L. J. (1989). *Sexual Visions: Images of Gender in Science and Medicine Between the Eighteenth and Twentieth Centuries.* Madison: University of Wisconsin Press.

Kevles, D. J. (1985). *In the Name of Eugenics: Genetics and Uses of Human Heredity.* Berkeley, CA, and London: University of California Press.

Kevles, D. J., and L. Hood. (1992). *The Code of Codes: Scientific and Social Issues in the Human Genome Project.* Cambridge, MA, and London: Harvard University Press.

Paul, D. B. (1995). *Controlling Human Heredity: 1865 to the Present.* Atlantic Highlands, NJ: Humanities Press.

———. (1998). *The Politics of Heredity: Essays on Eugenics, Biomedicine, and the Nature–Nurture Debate.* New York: SUNY Press.

Shreeve, J. (2004). *The Genome Wars: How Craig Venter Tried to Capture the Code of Life and Save the World.* New York: Alfred A. Knopf.

Integrating ELSI Concepts into Upper Division Science Courses

Anne Galbraith
UNIVERSITY OF WISCONSIN–LACROSSE

Science instructors often feel pressured to provide their majors with a certain amount of content in their courses. After all, many of these students will be moving on to higher level courses that depend on the foundation knowledge learned in prerequisite courses. In addition, students need a fair amount of content knowledge in order to successfully take exams to gain entrance into professional programs, exams such as the MCAT for medical school, the DAT for dental school, the PCAT for pharmacy school, and the Subject GRE for graduate school. Because of this, topics such as genetic testing and embryonic stem cells are often presented in class absent of their social or ethical contexts. Given that many science majors are destined to become future scientists and health professionals, it is imperative that they be given the chance to think about these technologies in terms of their effects on society, since it is, in fact, this society in which they will be living and serving.

I first became interested in the integration of ELSI issues into my courses in 1999, after my second semester of teaching at my current position at the University of Wisconsin–La Crosse. I had joined two other geneticists to team teach a graduate course called Advanced Genetics (see later description), and the first topic covered was eugenics. Even though my undergraduate through postdoctoral fellowship years encompassed over a decade, I had never heard about these atrocious activities that had occurred all in the name of using science to better society. When I polled one of my undergraduate courses that same semester, only two people out of about 110 had heard of the term eugenics.

In 2001, I had the distinct pleasure of being able to attend an NIH-sponsored Summer Faculty Institute at the Dartmouth College Ethics Institute

(Dartmouth College, 2006) to learn about teaching courses on the ethical, legal, and social implications of the Human Genome Project. It was a wonderful experience to connect with other faculty who were equally interested in integrating these concepts into their courses and/or develop entire new courses. I came home with a million ideas and vowed to modify three of my upper level courses: BIO306 Genetics, BIO466 Human Molecular Genetics, and BIO714 Advanced Genetics. The remainder of this chapter describes these courses and the topics that were introduced into them over the past five years as a result of my attending this institute at Dartmouth College.

The University of Wisconsin–La Crosse (UW-L) is a four-year comprehensive university that currently enrolls nearly 9,000 students, offering B.A., B.S., and M.S. degrees. Over the past several years, the credentials of the incoming freshman classes have been second only to the University of Wisconsin–Madison out of the thirteen public universities in the UW system. The numbers of our graduates who successfully apply to and complete graduate and professional programs in health-related fields continue to increase. The biology department, of which I am a faculty member, continues to increase its numbers of majors with each new freshman class, with over 700 declared biology majors currently. We offer a general Biology Major, as well as four concentrations: Biomedical Science, Cell and Molecular Biology, Aquatic Science, and Environmental Science. Although I teach several courses in addition to those described next, these are the three into which I have integrated ELSI concepts over the past five years.

BIO306 Genetics

This course meets three hours per week for lecture and has a two-hour lab. Undergraduates enrolled are predominantly sophomores and juniors. Genetics is required for all Biology majors and Biology minors at UW-L. It is also required for most preprofessional programs such as pre-Physical Therapy, pre-Radiation Therapy, and pre-Nursing. Therefore, this course has become increasingly popular. For example, in fall 2001, there were seventy students enrolled in this course and in fall 2006 there were nearly 160 enrolled, for the first time split into two lecture sections.

Due to sheer numbers and the pressure to include all content needed to produce successful graduates of this course, integrating ELSI concepts into this course was the most challenging. However, it has been simultaneously the most rewarding simply due to the high impact factor. There are four major ELSI concepts that have been integrated into this course: Eugenics, History of the Human Genome Project, Embryonic Stem Cells and Cloning, and Genetic Testing.

Eugenics. Early in the semester, I devote an entire lecture to the topic of eugenics. I begin by asking students how many of them have heard the term *eugenics.* Typically, out of a class of 120 students, no more than five raise their hand. I then launch into an introduction to Francis Galton and a description of the various organizations that were developed in the early 1900s to study the application of Mendelian genetics to humans (Gillham, 2001). I explain how information was provided to the public at state fairs, in religious sermons, using puppet shows, and in textbooks (Selden, 1999). I also include a description of the three major types of laws that were enacted as a result of the eugenics movement: miscegenation laws, immigration restrictions, and involuntary sterilization (Selden, 1999). Most of the graphics that I use in class have been borrowed from the Eugenics Archive site (Dolan DNA Learning Center, 2006), as genetics textbooks rarely, if ever, include this kind of information. The factoid that typically hits home the most is when I tell them that the last sterilization was done in 1979 (Lombardo, 2003). Even students who were born in the late 1980s realize that 1979 was not that long ago.

This topic is sensitive, so it is very important that the delivery be carefully executed. How the story is told to these students is just as important as what is told. I do not preach to the students. I simply tell the story, pointing out the incredible logic flaws, using sarcasm and dry humor to give them permission to laugh at how ridiculous this all was, while simultaneously maintaining an air of the seriousness of the implications of the use of scientific knowledge to shape public policy. I also go out of my way to impress upon the students that eugenicists were often noble people who were using the science as they understood it to improve society, and believed that they were doing the right thing.

However, the results, in hindsight, were appalling (Black, 2003). I then fast-forward to modern times, pointing out that we use science today to benefit society as well. We are noble, and try to do the right thing. However, 100 years from now, what we are doing now may be so obviously wrong.

History of the Human Genome Project. About a month after the topic on eugenics, I introduce the history of the Human Genome Project. I discuss the facts that not all scientists thought it was worth the money and effort, that given the state of the technology at the time that it would have been impossible to complete in a reasonable amount of time, that biologists had never before worked together on a huge project and this was "different" from how most life science had been done in the past (Roberts, 2001). I then point out the huge amount of information that has resulted from this project, including the sequencing and gene mining of an ever-growing number of model organisms, as well as improved technologies that are impacting our abilities to improve and personalize health care (NHGRI, 2006).

When I ask them who else may want to know this information, they recite the usual "insurance companies" answer. However, upon digging more deeply, they begin to realize that genetic information would also be useful for employers, mortgage companies, and even potential spouses! I then point out that we are only using our scientific knowledge and expertise to benefit society, to improve health care, to save lives, to treat disease. That is certainly a noble cause, but as hoped, several students always point out that this is what the eugenicists thought in the 1900s. Exactly!

Embryonic Stem Cells and Cloning. This topic is generally presented shortly after the topic on the Human Genome Project. Embryonic stem cells and cloning are part of a lecture in which other technologies are also introduced. The science behind these topics is described in our genetics textbook, so I use that as a resource (Hartwell et al., 2004). My goal is to provide the information about embryonic stem cells in an unbiased way, explaining the science and the controversy over the use of these cells. I then segue into the related topic of cloning, both therapeutic and reproductive, and the differences between the two.

One student in a course evaluation at the end of the semester commented on this topic in particular, even though it was taught at mid-semester, stating that s/he was glad that I had provided all sides since most science instructors simply teach these ideas as though they must agree with their use. "It was nice that you simply told us what was what, provided the sides of the arguments for and against, and let us make up our own minds" (2005).

Genetic Testing. One of the labs for this genetics course is called RFLPs, Pedigrees, and Genetic Testing. In this exercise, students are genetic researchers trying to develop a diagnostic test for a genetic disorder. The test they are performing will use restriction-length polymorphisms

(RFLPs) to look for differences in a person's DNA that correlates with the genetic disorder. Plasmids carrying different-sized DNA inserts are used as the samples. In this way the same restriction enzymes can be used to generate products of different sizes. By combining the plasmids, "heterozygous individuals" can be created, illustrating the concept of carriers of a disease. This lab was developed in collaboration with my geneticist colleague in the biology department, Dr. Mike Abler.

I take this opportunity not only to teach them how simple it is to test people genetically for disorders, but also to provoke discussion (no more than twenty students are enrolled in each lab section) using questions such as the following:

- Some people have argued that performing genetic tests is wrong for several reasons. How do you feel about these arguments? Do you think that you would want to have these tests done on you and your spouse before deciding to have children?
- Genetic tests are a powerful way of predicting who is at increased risk of developing a genetic disease. Insurance companies could also use these results against a person, charging them more for health insurance because they have a preexisting condition. Is this right or should healthy people be forced to pay higher premiums to cover genetic risk?
- Genetic testing does not cure a disease; it only tells you who is more likely to get a disease. Would you want to be told at age twenty that when you were fifty you were going to get Huntington disease, a condition for which there is presently no cure?

During these discussions, I often find that there are students in the class willing to contribute that they have a genetic disorder in their family. Often, these students are also willing to provide a personal account of how this genetic disorder, and genetic testing, has impacted their lives.

BIO466 Human Molecular Genetics

This course meets three hours per week for lecture and has an optional three-hour lab. It is an elective course for the general Biology Major, as well as for the Biomedical Science and Cell and Molecular Biology Concentrations. Many of the students who take this course plan to attend a professional or graduate program in the future. Most of these students are seniors, although there are often two to five graduate students enrolled as well. The only prerequisite course is BIO306 Genetics. I first offered this course in fall 1999 to twenty-one students. There were thirty-nine students enrolled for fall 2006. Clearly the impact factor for this course continues to rise. Given the focus of the course on humans, ELSI concepts, including eugenics, have been infused into the course at many points. However, the concept that I focus on here is the "human-ness" of genetic disorders.

For this topic, I borrowed an idea from Koehler and Hawley (1999). The students are randomly assigned a genetic disorder topic on the first day of class. The students are required to write a paper that answers several scientific questions about the disorder within a human context of interpersonal and family relationships and/or social and ethical issues. All the scientific questions can be answered with information from the textbook and class lectures. Students are encouraged to use other sources such as web sites and support groups to help understand the personal, emotional, and social consequences of their assigned disorder. Students are then required to use the scenario to write a fictional account of the lives of the people involved in the situation that "can be understood by their grandmother." Creative approaches to integrating the science into a short-story-type format are expected, but regardless of the format, the information must be written up as a detailed account of the lives of the fictional people trying to cope with their genetic condition.

I admit to my students that I could just ask them to write "yet another term paper," but I tell them that as seniors and graduate students, they'd better know how to do that by now! This paper forces students to write about science for the nonscientist, a skill that will serve them well in their futures as resources for the nonscience public. Although I initially borrowed several of the published scenarios that were presented in the Koehler and Hawley article (1999), I have also written several of my own over the years. One example of a scenario that I wrote is provided here:

Janice and Erik Simpson are the healthy parents of five-year-old Stephen. At a young age, Stephen was diagnosed with DMD (Duchenne muscular dystrophy). What is DMD and what are the symptoms? How is this disorder inherited and what is the defect? Janice and Erik are discussing the possibility of having another child. However, they are concerned about going through the emotional and financial turmoil of having a second child with DMD. What types of emotional and financial problems has this family probably faced since the diagnosis? What is Stephen's prognosis? Stephen's parents are seeing a genetic counselor, Dr. Laurie Peters. What are the chances of Janice and Erik having a daughter with DMD? A son? Why are the chances higher for having a son with DMD? How should Dr. Peters advise the Simpsons? Write from the perspective of Janice, Erik, Stephen, or Dr. Peters.

I have found over the years that students who do the best in their classes tend to have a harder time doing a good job on this assignment. In anonymous course evaluations, students often remark that this was one of the hardest scientific papers they had ever written, but it was also the most enjoyable. With well over 150 students having taken the course, no students have ever said that they did not enjoy the assignment or failed to see its usefulness to their education.

BIO714 Advanced Genetics

This course meets three hours per week. It is a master's-level course that can be taken by undergraduates with the permission of an instructor. There are currently three instructors for this course, including my geneticist colleague and myself from the biology department, and another geneticist, Dr. Marc Rott, from the microbiology department. Many of the undergraduates who take this course plan to attend graduate school, although there are a few pre-Medical students who have taken the course. The prerequisite course is either BIO306 Genetics, or the microbiology equivalent, MIC416 Microbial Genetics. The course is offered every other spring, and although it initially enrolled only five students in 1999 when I first became a part of it, our last offering was to fourteen students. This course is designed to teach students how to read primary literature critically. The topics that we discuss vary with the semester, as topics chosen are based on student and instructor interest. However, there are two major ELSI concepts that are always covered in this course: eugenics, and scientific responsibility. As I have already discussed the topic of eugenics earlier, I focus on the topic of scientific responsibility here.

Several short readings are assigned for this first topic of the semester. For example, an article by Beckwith (1993) addresses the public image of scientists. A letter by former editor of the journal *Science*, Daniel Koshland, Jr. (1989), predicts that the sequencing of the human genome will rid society of a series of physical and mental illnesses, and that the benefits far outweigh the risks. An article by Allen (2001) laments "the false promise of a technological fix."

We begin the discussion with the Beckwith (1993) article by asking students to describe a typical scientist and the influences that have helped shape that image for them. We then dive into a discussion of all three papers: Should scientists be held responsible to society for their science? Should scientists be more careful when instructing the nonscience public about the power and benefits of new technologies? Are scientists responsible to society if they do overestimate these benefits?

This is always a fascinating conversation, as students typically differ in their views and are quite ready to argue. For example, some students think that scientists should simply do their jobs to progress scientific knowledge, and that is it. Any influence of their science on society or public policy is not their concern. Other students think that scientists have an obligation to think about the impact of their potential discoveries on society before they even perform the experiments. Of course, the opinions of most students fall somewhere in between these two extremes. Regardless of the direction that is taken by this conversation, it is an important conversation for future scientists and health professionals to have.

Conclusions

After participating in a Summer Faculty Institute at Dartmouth College in 2001, I learned more about the importance of developing curricula on the ethical, legal, and social implications of modern science. Over the past five years, I have integrated some of these topics into three of my science courses. Anonymous course evaluations have consistently been overwhelmingly positive when asked specifically about those topics.

As with most lessons, modifications have been made over the years to retain the most significant information and incorporate new information. Despite the fact that a lot has changed in the field of genetics over the past five years, the need for students to discuss the ethical, legal, and social implications of new technologies has only increased.

References

Allen, Garland (2001). Is a New Eugenics Afoot? *Science* 294: 59–61.

Beckwith, Jon (1993). A Historical View of Social Responsibility in Genetics. *Bioscience* 43: 327–333.

Black, Edwin (2003). *War Against the Weak*. New York: Four Walls Eight Windows.

Dartmouth College (2006). http://www.dartmouth.edu/~ethics [2006, June 9].

Dolan DNA Learning Center (2006). http://www.eugenicsarchive.org/eugenics [2006, June 9].

Gillham, Nicholas (2001). Sir Francis Galton and the Birth of Eugenics. *Annual Review of Genetics* 35: 83–101.

Hartwell, Leland, Leroy Hood, Michael Goldberg, Ann Reynolds, Lee Silver, and Ruth Veres (2004). *Genetics from Genes to Genomes.* New York: McGraw-Hill.

Koehler, Kara, and R. Scott Hawley (1999). Tales from the Front Lines: The Creative Essay as a Tool for Teaching Genetics. *Genetics* 152: 1229–1240.

Koshland, Daniel, Jr. (1989). Sequences and Consequences of the Human Genome. *Science* 246: 189.

Lombardo, Paul (2003). Taking Eugenics Seriously: Three Generations of ??? Are Enough? *Florida State University Law Review* 30: 191–218.

NHGRI (National Human Genome Research Institute) (2006). http://www.genome.gov [2006, June 9].

Roberts, Leslie (2001). Controversial from the Start. *Science* 291: 1182–1188.

Selden, Steven (1999). *Inheriting Shame: The Story of Eugenics and Racism in America.* New York: Teachers College Press.

Glossary of Terms

Advanced directive—a written document that indicates choices about medical treatment.

Allele—an alternative form of a gene located at a specific position on a specific chromosome. In humans and other organisms having two copies of each chromosome, each gene exists in at least two version, or alleles.

Biotechnology—the alteration of cells or biochemicals with a specific application, including monoclonal antibody technology, recombinant DNA technology, transgenic technology, and knockout and knockin technologies.

Blastomere—one of the cells that make up the early embryo.

Chimera—an animal in which cells from other animals have been inserted early in embryonic development.

Chromosomes—units of genetic material, composed of DNA and protein structures, across which the three billion pairs of nucleotide letters of the genome are distributed. The human genome has forty-six chromosomes.

Cloning—the process of making genetically identical copies.

Coding strand—the side of the double helix for a particular gene from which RNA is not transcribed.

DNA—(deoxyribonucleic acid) the long molecule composed of nucleotides that constitute the molecular basis of heredity. The nucleotides are lined up on two connecting complementary strands that twist along their length to form a double-helix structure.

ELSI—the Ethical, Legal, and Social Implications of the Human Genome Project.

Embryonic stem cells—also known as human pluripotent stem cell; self-replicating, known to develop into cells and tissues of the three primary germ layers.

Eugenics—the control of individual reproductive choices to achieve a societal goal.

Gene—a sequence of DNA letters (nucleotides) that codes for the proteins and other chemical elements that make up the structures and direct the functions of a biological organism.

Gene pool—all genes in a population.

Gene therapy—replacing a malfunctioning gene to alleviate symptoms.

Genetic determinism—the idea that an inherited trait cannot be modified.

Genetic markers—a gene or DNA sequence having a known location on a chromosome and associated with a particular gene or trait.

Genetic testing—tests done for clinical genetic purposes such as disease or prediction of drug responses.

Genetically modified organisms (GMOs)—a plant, animal, or microorganism produced or changed by genetic engineering.

Genome—all the genetic material in the cells of a particular type of organism.

Genotype—the allele combinations in an individual that cause a particular trait or disorder.

Germline gene therapy—genetic alterations of gametes or fertilized ova, which perpetuate the change throughout the organism and transmit it to future generations.

Haplotype—a series of known DNA sequences linked on a chromosome.

Human Genome Project (HGP)—the thirteen-year-long project (1990–2003) coordinated by the U.S. Department of Energy and the National Institutes of Health to sequence the three billion nucleotides contained in the human genome.

Health Insurance Portability and Accountability Act of 1996 ("HIPAA")—a law designed to provide privacy standards to protect patients' medical records and other health information provided to health plans, doctors, hospitals, and other health care providers.

IVF—in vitro fertilization, a technique with which egg cells are fertilized by sperm outside the woman's womb.

IVM—in vitro maturation, a still-undeveloped technology for maturing eggs (oocytes) outside the woman's body to the point where they are ready for fertilization.

Junk DNA—sequences of DNA in the gemone that do not code for proteins or other regulatory products. It is not established, however, that all such DNA has no function.

Mendelian traits—genetic traits or diseases caused by a variation or mutation in a single gene and usually inherited in dominant or recessive patterns.

Mutagen—a substance that changes a DNA base.

Mutant—an allele that differs from the normal or most common allele, altering the phenotype.

Mutation—a change in a protein encoding gene that has an effect on the phenotype.

NIH—National Institutes of Health.

NTD—neural tube defect, a major birth defect caused by abnormal development of the neural tube, the embryonic structure that gives rise to the central nervous system.

Phenotype—the invisible or detectable characteristics of an organism shaped by the organism's environment.

Preimplantation genetic diagnosis (PGD)—removing a cell from an eight-celled embryo and testing it for a disease-causing gene or chromosomal imbalance, to decide whether the remaining embryo should be implanted in the uterus to continue development.

Recessive gene—refers to an allele that causes a phenotype only when both chromosomes carry that allele.

Single-nucleotide polymorphisms, SNPs—a variation in DNA sequence occurring whenever one or more nucleotides—A,T, C, or G—in the genome differs from one individual to another.

Social Darwinism—a social theory that states that the level a person rises to in society and wealth is determined by the person's genetic background.

Somatic cell gene—a technique for altering genes present in the somatic cells (or gene body) of an individual (therapy affecting sex cells). Unless the changes accidentally transfer, they are not capable of being passed on to subsequent generations.

Stem cell—relatively undifferentiated cells that retain the ability to renew themselves indefinitely through cell division and to differentiate into a wide range of specialized cell types.

Tay–Sachs disease—a neurodegenerative disorder that leads to death in early childhood. It is more prevalent in eastern European Jewish populations than in others.

Contributors

Beecher-Monas, Erica
Professor
Wayne State University Law School
Detroit, Michigan

Corrette-Bennett, Joshua
Assistant Professor, Department of Biology
Westminster College
New Wilmington, Pennsylvania

Donovan, Aine
Executive Director, Ethics Institute, and Research Associate Professor
Dartmouth College
Hanover, New Hampshire

Eggleston, Ben
Professor, Department of Philosophy
University of Kansas
Lawrence, Kansas

Eubanks, Sonja R.
Associate Director, Counseling Program
University of North Carolina–Greensboro
Greensboro, North Carolina

Galbraith, Anne M.
Associate Professor, Department of Biology
University of Wisconsin
La Crosse, Wisconsin

Gladstone, Julia Alpert
Associate Professor, Department of Law
Bryant University
Smithfield, Rhode Island

Green, Ronald M.
Faculty Director, Ethics Institute
Professor, Department of Religion
Dartmouth College
Hanover, New Hampshire

Hicock, Bethany
Associate Professor, Department of English
Westminster College
New Wilmington, Pennsylvania

Jackson, Myles W.
Dibner Family Chair of the History of Science and Technology
Polytechnic University
New York, New York

McConnell, Terrance
Professor, Department of Philosophy
University of North Carolina–Greensboro
Greensboro, North Carolina

Sander-Staudt, Maureen
Assistant Professor, Department of Philosophy
Arizona State University
Tempe, Arizona

Segady, Tom
Professor, Department of Sociology
Stephen F. Austin University
Nacogdoches, Texas

Stober, Spencer S.
Associate Professor, Department of Science and Math
Alvernia College
Reading, Pennsylvania

Werth, Alexander
Professor, Department of Biology
Hampden-Sydney College
Hampden-Sydney, Virginia

Yarri, Donna
Associate Professor, Departments of Theology and Humanities
Alvernia College
Reading, Pennsylvania

Index